TREE DISEASES OF EASTERN CANADA

Editor
D.T. Myren

Associate Editors
G. Laflamme, P. Singh, L.P. Magasi, and D. Lachance

Published by

Natural Resources Canada
Canadian Forest Service
Science and Sustainable Development Directorate

Ottawa, 1994

Canada Communication Group Publishing
Groupe Communication Canada Édition

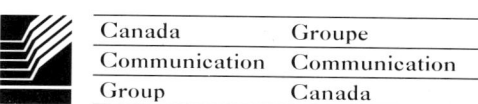

Available in Canada through your local bookstore
or by mail from Canada Communication Group – Publishing,
Ottawa, Canada K1A 0S9

Catalogue No. Fo42-186/1994E
ISBN 0-660-14936-2

Production: Paula Irving
Text editor: Marla Sheffer
Cover design: Steven Blakeney

Cette publication est aussi disponible en français
sous le titre *Maladies des arbres de l'est du Canada*

Canadian Cataloguing in Publication Data

Main entry under title:

Tree diseases of eastern Canada

Issued also in French under title: Maladies des arbres de l'est du Canada.
ISBN 0-660-14936-2
DSS cat. no. Fo42-186/1994E

1. Trees — Diseases and pests — Canada, Eastern.
2. Plant diseases — Canada, Eastern. I. Myren, D.T.
II. Laflamme, G. III. Canada. Forestry Canada.
Science and Sustainable Development Directorate.

SB605.C3T73 1994 634.9'6'0971 C93-099563-5

Forestry Canada is now called the Canadian Forest Service and forms part of a new federal department entitled Natural Resources Canada.

Printed on recycled paper

Printed on alkaline permanent paper

PRINTED IN CANADA

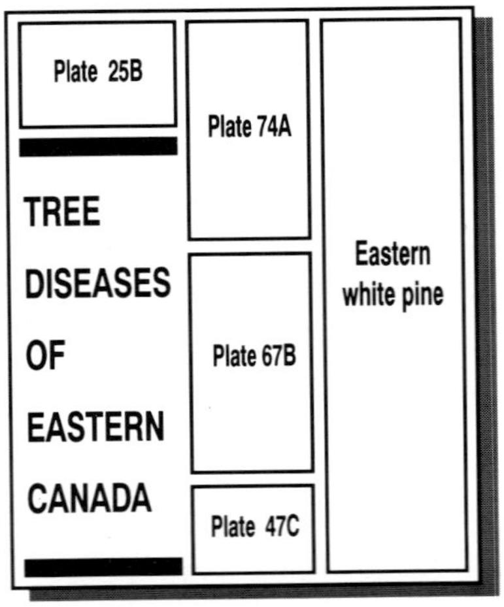

Cover: Plate 25B. *Davisomycella ampla* (J. Davis) Darker.
Plate 74A. *Armillaria mellea* complex.
Plate 67B. *Cronartium ribicola* J.C. Fischer.
Plate 47C. *Gremmeniella abietina* (Lagerb.) Morelet.
Eastern white pine (*Pinus strobus* L.).

Contents

Diseases of Stems and Branches

Diseases of Cones

Damage Caused by Animals and Insects

Damage Caused by Abiotic Agents

Contributors

René Cauchon
Laurentian Forestry Centre
Forestry Canada, Quebec Region
P.O. Box 3800
1055 du P.E.P.S. Street
Sainte-Foy, Quebec
G1V 4C7

Henry L. Gross
Great Lakes Forestry Centre
Forestry Canada, Ontario Region
P.O. Box 490
1219 Queen Street East
Sault Ste. Marie, Ontario
P6A 5M7

Denis Lachance
Laurentian Forestry Centre
Forestry Canada, Quebec Region
P.O. Box 3800
1055 du P.E.P.S. Street
Sainte-Foy, Quebec
G1V 4C7

Gaston Laflamme
Laurentian Forestry Centre
Forestry Canada, Quebec Region
P.O. Box 3800
1055 du P.E.P.S. Street
Sainte-Foy, Quebec
G1V 4C7

André Lavallée
Laurentian Forestry Centre
Forestry Canada, Quebec Region
P.O. Box 3800
1055 du P.E.P.S. Street
Sainte-Foy, Quebec
G1V 4C7

Laszlo P. Magasi
Maritimes Forestry Centre
Forestry Canada, Maritimes Region
P.O. Box 4000
College Hill
Fredericton, New Brunswick
E3B 5P7

Donald T. Myren
Great Lakes Forestry Centre
Forestry Canada, Ontario Region
P.O. Box 490
1219 Queen Street East
Sault Ste. Marie, Ontario
P6A 5M7

Guillemond B. Ouellette
Laurentian Forestry Centre
Forestry Canada, Quebec Region
P.O. Box 3800
1055 du P.E.P.S. Street
Sainte-Foy, Quebec
G1V 4C7

Pritam Singh
Forestry Canada
Place Vincent Massey
20th Floor
351 St. Joseph Boulevard
Hull, Quebec
K1A 1G5

Acknowledgments

The editors would like to acknowledge their indebtedness to the field staff of the Forest Insect and Disease Survey who have provided much of the field material, given freely of their observations, and made available many special collections for our examination. We also wish to express our appreciation to our collegues for sharing their knowledge with us and for their many helpful comments. We would particularly like to thank G.C. Carew, André Carpentier, L. DiPasquale, E.B. Dorworth, C. Handfield, J.E. Hardy, P. Jakibchuk, A. Jones, G. Katagis, and L. Mamanko for helping in so many ways.

Introduction

Plant pathology is the study of plant diseases, their causes, and their control. Plant diseases may be caused by biological organisms, such as fungi and bacteria, or nonbiological agents, such as herbicides, air pollution, and drought. Forest pathology is a specialized area of plant pathology that is concerned with the diseases of trees. In 1874, the German forester Robert Hartig established that the fungi seen on tree trunks were in fact the fruiting bodies of fungal threads hidden within the decayed wood of the tree. This concept, revolutionary in its day, is recognized as the starting point of forest pathology as a science.

Forest pathologists are concerned with volume loss and mortality of forest trees. These are caused by three major factors: diseases, insects, and fire. In the forest, fungi are considered to be the most significant disease-causing agents. Diseases outrank both insects and fire as a cause of volume loss, whereas insects are considered to be the leading cause of tree mortality. Losses due to disease and even to insects are insidious and often not observed, whereas losses due to fire are obvious and spectacular.

Forest pathologists are also concerned with the aesthetic and economic impact of tree diseases. The high cost of growing trees for the Christmas tree market, for example, makes losses due to disease serious in these plantations. Seed orchards and windbreaks, which are also of considerable value, can be attacked and damaged by pathogenic fungi. Losses of ornamental trees result in a reduction in aesthetic value as well as expensive removal and replacement costs.

Before disease control measures can be initiated, the problem must be understood. Control measures might not even be required. Identification of the cause of tree diseases is a specialized area, but many of the more frequently encountered diseases can be recognized readily if adequate descriptive information is available. It is our hope that the descriptions and photographs presented in this book will enable the reader to identify and select a proper course of action in dealing with the majority of the more common tree diseases.

Topics included in each description are hosts, distribution, effects on hosts, identifying features, life history, control, additional information, and selected bibliography. Hosts and distributions of the causal agents pertain to eastern Canada, although many of them are also found in western Canada and the United States. Identifying features include macroscopic characters and those that could be observed with a 10× microscopic lens. Life histories are presented in their entirety when known; however, some diseases included are not fully understood, and consequently the life history of the causative organism is very brief.

Discussions of control measures have purposely omitted specific fungicides because of the changes that occur in their availability and registration, as well as the constant introduction of products with greater efficacy. It is suggested that those planning a control operation in which pesticides will be used contact an agricultural or forest agency in their area for current recommendations.

Additional information briefly discusses similar pests, methods of sampling, and other facts that may be of interest. It is hoped that the reader will follow our advice regarding selection of samples should it be necessary to send material to specialists for examination. The selected bibliography provides sources of additional information, and references to other works on a particular disease will be found in most of these.

The diseases selected for coverage were chosen after consultation among forest mycologists and pathologists of the four regional forestry centers of Forestry Canada (Newfoundland Forestry Centre, Forestry Canada—Maritimes, Laurentian Forestry Centre, and Great Lakes Forestry Centre), examination of the historical records of the Forest Insect and Disease Survey units located at these centers, and a review of the scientific literature.

Causal Agents of Tree Diseases

The factors that cause diseases in trees are often referred to as agents. These agents can be living organisms, such as fungi, bacteria, and viruses, which can cause actual infection of the tree. There are also nonliving agents, which cause injuries at sites that may later serve as points where infection can be initiated. In some cases, the injuries caused by both living and nonliving agents may be so severe that the affected plants die. Possibly the best way to classify these agents is according to their infectiousness: those that can multiply within the host are termed "infectious," and those that do not are called "noninfectious." Infectious agents produce some form that enables them to spread and infect other hosts.

Man, animals, and weather are all noninfectious agents. We selected the most important or common agents that can cause damage for individual coverage in the text.

Infectious diseases of trees are primarily caused by fungi. Fungi are simple plants that lack chlorophyll and that colonize their substrate by the development of microscopic, thread-like structures called hyphae or hyphal strands. Because fungi have no chlorophyll, photosynthesis cannot occur; hence, fungi obtain food by enzymatic action from the material upon or in which they are growing.

Fungi reproduce by spores, which function much like seeds but are so small they can be seen only with the aid of a microscope. Spores are primarily spread by wind, rain splash, insects, and movement of soil and infected plant material. When a spore lands on a suitable site it germinates, producing a germ tube that develops into a hyphal strand as the fungus develops. These strands grow, branch, and, in favorable conditions, spread rapidly.

Many fungi are beneficial. Some function to break down dead organic matter; some (yeasts) are used in baking and brewing; some (molds) are used in cheese and antibiotic production; and others (mushrooms) produce edible fruiting bodies.

There are also a number of fungi that attack and damage living plants. Infectious fungi or the diseases they cause can be categorized as fluctuating or nonfluctuating. Those that fluctuate vary in their presence and severity from year to year. Leaf spots are good examples of fluctuating diseases. Nonfluctuating diseases are those that persist once they are established. Cankers on stems or branches are good examples: they persist until the host or infected part perishes. Some fungi even remain active on a dead host.

The diseases we have selected for coverage can be grouped into the following categories: rusts, decays, wilts, cankers, needle casts, and anthracnose.

The **rust diseases** are a very interesting group, because many of them require two different living hosts to complete a rather complex life cycle. The host of lesser economic importance is usually called the alternate host. Rust fungi get their common name from the rusty orange color of their spores during at least one of their fruiting stages. A complete life cycle has five separate fruiting stages. Many rust fungi do not go through all five of the stages, and some are able to complete their life cycle on a single host. Rust fungi derive their nourishment from living plant cells, so that they die if their hosts die. Tree rusts have representatives in both the fluctuating and nonfluctuating disease groups.

The **wood decay fungi** are particularly insidious disease agents, as much of their activity occurs inside their hosts without any obvious external symptoms. Some of these fungi are found to cause decay of the roots, and others are confined primarily to the stem. Once the decay fungus produces a fruiting structure, indicating its presence in the tree, the decay may be well advanced.

There are two major groups of decay fungi: white rots decompose all of the wood components, and brown rots decompose just the cellulose, leaving the lignin. It is estimated that there are about 1700 decay fungi, of which about 6% are brown rots.

Wilt diseases are caused by fungi that invade the vascular system of the host. They interfere with the translocation of fluids within the tree, resulting in a reduced flow of water to the leaves and subsequent wilting. The mechanism by which the fungus causes the interruption of fluid movement is not completely understood, but it probably involves physical obstruction and toxin production by the fungus and the development of structures by the host that tend to plug water-carrying vessels. Trees infected by wilt fungi often have a solid or dotted pattern of color in the outer sapwood when seen in cross section. Wilt diseases are nonfluctuating, eventually resulting in mortality.

Canker fungi cause distortions of the trunk or branches of infected trees and are found on both hardwoods and conifers. Damage ranges from volume loss of varying degrees to death. Cankered trees under stress from wind or heavy ice and snow accumulation often break at the point of cankering. Canker fungi frequently invade their hosts through branch stubs or wounds. Once established, the fungus kills the bark, often resulting in a characteristic pattern or color as the host responds to the invasion. Canker fungi fruit on the host, and their spores can be liberated whenever temperature and moisture requirements are met.

Needle cast fungi are common on conifer needles, and many are capable of causing premature defoliation. Infection usually occurs on needles of the current year, and the reproductive stage of the fungus may occur at the end of that season or up to 2 years later. The reproductive structure is usually black and may be circular, oval, or elongate. It may cover up to the entire length of the needle. The fruiting structures of the needle cast fungi can be seen with the unaided eye.

Most of the needle cast fungi do not cause a serious problem, but several are capable of causing significant damage to young trees and to trees in forest nurseries. Needle cast is a fluctuating disease.

Anthracnose is a disease of hardwood foliage caused by fungi that spend the winter on fallen infected leaves or in the twigs. In the spring, spores discharged by the fungi infect the new leaves. These fungi are capable of causing considerable destruction of leaf tissue as well as premature defoliation. Anthracnose is a fluctuating disease and is not likely to be fatal.

Not all of the organisms covered in this text fit into the above groups. For example, fire blight is caused by a bacterium. Bacteria are very small, single-celled organisms that are responsible for a lot of soft rot and can kill tissue of living hosts. They are often disseminated by insects. Dwarf mistletoe is a parasitic seed plant that infects some conifer species. It is disseminated by a forceful ejection of its seeds and possible incidental transport by birds and small mammals.

Collection of Disease Specimens

Selection of good samples of tree diseases is the key that allows the specialist to identify a causal factor. Information describing many aspects of the site is also important. The recommendations provided here have been prepared with the forester in mind but should serve as guidelines for anyone who has occasion to ship samples of tree diseases.

Collection procedures

The following points are important to consider when collecting specimens of tree pathogens:

1. As large a sample as possible of well-developed, representative host material bearing fruiting bodies of the fungus should be secured. (Some fungal structures are so small they are almost invisible to the naked eye.)

2. Each collection should contain only one type of damage from a single diseased tree, but a number of collections can be shipped together.

3. Each collection should be placed in a paper bag, tube, or container together with a note containing observations and comments that should have been completed in the field at the time the collection was made. The note should be placed in a small plastic bag if the sample is moist, or the bag containing the collection can be coded and referred to in a separate letter.

4. Samples of twigs, branches, or roots should be 10–15 cm long, and each should include the margin between living and diseased tissue.

5. Collections of leaves, ferns, small plants, and plant branchlets should be pressed between pages of a newspaper or magazine or between cardboard when collected, and shipped, without bending, in a protective container.

6. Large fruiting bodies should be dried in the open air, and soft, fleshy fruiting bodies should be dried quickly and completely in warm, dry air. The color, odor, and size of fruiting bodies and points of attachment to the host should be noted at the time of collection, as many of these characteristics are lost after drying.

7. Samples should not be wrapped in cellophane or plastic wrap, because the high humidity that results induces the development of contaminating fungi and bacteria, which make the detection of the true pathogen difficult.

8. In the case of heteroecious rusts (that is, rusts that produce their spore forms on two different hosts), samples should be collected from both hosts whenever possible.

Information to include

Although it is recognized that all the information listed below may not be available, an effort should be made to provide as much of it as possible to increase the probability of a successful identification.

1. The full name and address of the person(s) to whom the answer should be sent and the location where the sample was collected should be included.

2. The tree species should be identified, and the following information should be provided: diameter at breast height; height (estimate); whether living or dead (if possible); and the part of the tree affected — foliage, flowers, fruit, twigs, branches, stem, butt, or roots.

3. The status of the disease — decreasing, increasing, or constant — should be given.

4. The number of trees affected should be counted or estimated. If more than one tree is affected, it should be noted whether the trees are scattered, how many are involved, whether it is a plantation, whether the disease is confined to only a part of the area, and so forth.

5. Contributing factors should be listed: for example, site, poor drainage, mechanical damage just above the ground line, recent construction (within the last 5 years), etc.

6. Remarks on disease symptoms, such as yellowing, wilting, cankers, callus, flagging, girdling, resin-flow, dieback, dying, and dead top, should be included. Signs of the fungus, such as fruiting bodies, should also be mentioned. Other facts that might aid in diagnosing and appraising the situation, such as the possibility that salt from winter snow clearing operations could be involved or the possibility of urine damage from animals, should also be included.

1. Anthracnose
Apiognomonia errabunda (Roberge) Höhnel
Plate 1

Hosts: Mainly maple and oak; occasionally ash, beech, and sycamore.

Distribution: Common throughout eastern Canada.

Effects on hosts: Affected trees are not usually killed but may be weakened, thus becoming more vulnerable to damage by other agents. Infected leaves often fall prematurely.

Identifying features: Damage is characterized by the development of irregular areas of dead tissue on the leaves of the host. These areas vary in size, sometimes killing the entire leaf, including the veins, and may extend down the petiole into the young twigs.

Life history: The anthracnose fungi overwinter on the infected leaves or twigs, discharging spores the following spring when conditions are right for infection. Once infection has occurred, dead tissue develops, indicating the presence of the disease. Spores produced on the dead tissue are readily spread by rain splash and wind, resulting in the intensification of the disease.

Control: Raking and destroying infected leaves in the fall help control the fungus by reducing its overwintering population. Fungicides provide good control and are recommended for use after a tree has suffered a year of severe damage. They are not recommended for routine use, because anthracnose incidence fluctuates from year to year — in some years, the disease is almost nonexistent.

Additional information: The imperfect state of *A. errabunda* is *Discula umbrinella* (Berk. & Broome) Morelet, but it is known by many synonyms. Two other fungi that also cause anthracnose on maple are *Aureobasidium apocryptum* (Ell. & Ev.) Hermanides-Nijhof and *Cryptodiaporthe hystrix* (Tode : Fr.) Petrak. The imperfect state of the latter is *Diplodina acerina* (Pass.) B. Sutton. In early literature, these fungi were placed in the genus *Gloeosporium.*

Anthracnose can be confused with leaf damage caused by nonliving agents, such as hot, dry winds (see Section 93). Laboratory examination is often necessary to determine the true cause of the problem. Samples submitted to a diagnostic facility for identification should be well pressed.

The species *Apiognomonia errabunda* is very broad in the concept proposed by von Arx, which is open to question. Many mycologists feel that some of the species brought under this name (particularly *Apiognomonia veneta* (Sacc. & Speg.) Höhnel and *Apiognomonia quercina* (Kleb.) Höhnel) are distinguishable and should be recognized as such. The broad concept of the species is used in this instance, but students, in particular, are cautioned about its use.

Selected bibliography

Arx, J.A. von. 1970. A revision of the fungi classified as *Gloeosporium.* 2nd ed. Bibliotheca Mycologica, Band 24, J. Cramer, Lehre, Germany. 203 p.

Barr, M.E. 1978. The Diaporthales in North America, with emphasis on *Gnomonia* and its segregates. Mycol. Mem. No. 7. J. Cramer, Lehre, Germany. 232 p.

Boyce, J.S. 1961. Forest pathology. 3rd ed. McGraw-Hill Book Co., New York, NY. 572 p.

Prepared by D.T. Myren.

Plate 1

A. Anthracnose of oak caused by *Apiognomonia errabunda.*

B. Anthracnose of sycamore caused by *Apiognomonia errabunda.*

C. Defoliation of sycamore shoots due to anthracnose caused by *Apiognomonia errabunda.*

D. Anthracnose of maple caused by *Aureobasidium apocryptum.*

E. Maple leaves with anthracnose typical of *Aureobasidium apocryptum.*

F. An aerial view of severe anthracnose of maple caused by *Aureobasidium apocryptum.*

A

B

C

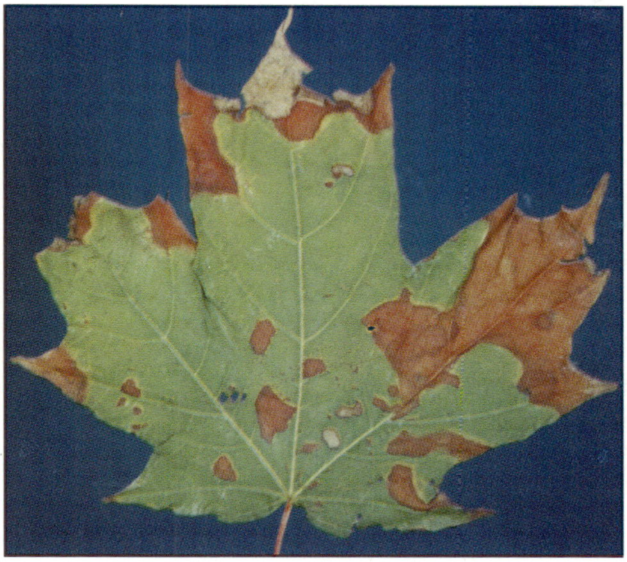

D

E

F

2. Cytospora dieback
Valsa friesii (Duby) Fuckel
Plate 2

Hosts: Mainly balsam fir; rarely red pine and Scots pine.

Distribution: Reported on balsam fir in all provinces of eastern Canada except Newfoundland.

Effects on hosts: This fungus is a common inhabitant of dead needles. It is suspected of being a needle pathogen and may have the potential to kill young shoots. The dieback with which *V. friesii* is associated does not kill its host but is known to deform small trees. Although in some areas up to 25% of the new shoots have been observed to have dieback symptoms, infection levels of 10% or less are more typical.

Identifying features: The medium-brown-colored needles on recently killed shoots are typical and evident during much of the year. The fungus fruits on the upper surface of the dead needles as small, erumpent, black specks. Careful examination reveals a black spherical structure under the needle surface just below these specks. Spores are produced within this body. Fruiting also occurs on the twigs but may be obscured by saprophytic fungi. Similar symptoms have been noted for which a cause could not be established. On twigs, the fruiting structure appears as a very small mound with a hole at the top through which the spores exude. Fruiting bodies have also been found embedded in the bark at crotches located at the base of the dead shoots. Cankers on larger branches have been reported but were not found on any of the *V. friesii*-infected material collected in Ontario.

Life history: Information on the life cycle of *V. friesii* is meager. Samples collected in late summer or early fall and the spring show mature fruiting bodies. Spores are liberated from the fruiting structure, possibly in tendrils, and are presumably spread by rain splash or wind. Infection probably occurs in the spring, possibly even the fall, when mature spores are present, but there are no field or laboratory studies to support these suppositions. The dead shoots can be seen in early summer, but the exact time this symptom becomes apparent is not known.

Control: Damage seems to be confined to small balsam fir in the forest, and no control measures have been deemed necessary. Removal of infected material certainly limits the spread of the fungus but is unwarranted in most situations. Branches should not be pruned in wet weather, and pruning tools should be swabbed with denatured alcohol between cuts, should one choose to attempt control.

Additional information: The imperfect state of this fungus, *Cytospora pinastri* (Fr. : Fr.), occurs commonly. Some authors have placed this state into a different species, calling it *C. friesii* Sacc. Both names should be used in literature searches.

Samples for diagnosis should include dead branches and brown needles with fruiting structures, if possible.

Two related fungi, possibly with somewhat more potential as pathogens, are *Valsa pini* (Alb. & Schwein.) Fr. and *Valsa abietis* (Fr. : Fr.) Fr. *Valsa pini* causes cankers on pine, and *V. abietis* is asso-

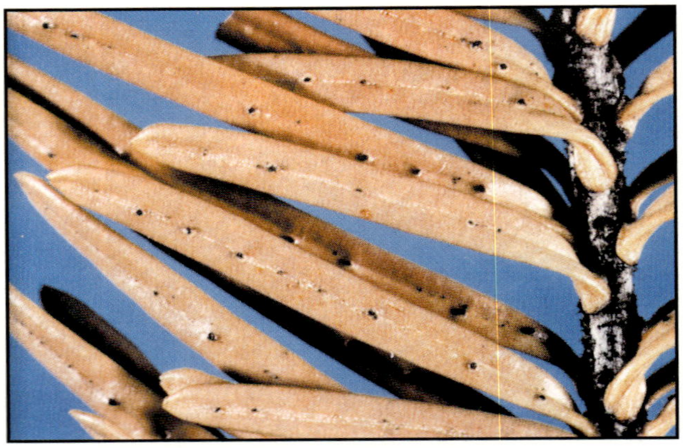

Plate 2

Balsam fir needles affected by cytospora dieback with fruiting bodies of the causal fungus *Valsa friesii.*

ciated with branch dieback of balsam fir. Both fungi have imperfect states in the genus *Cytospora*. Because *V. abietis* is found on balsam fir, it is often confused with *V. friesii*.

Selected bibliography

Grove, W.B. 1935. British stem and leaf fungi. Vol. 1. Cambridge University Press, London. 488 p.

Kobayashi, T. 1970. Taxonomic studies of Japanese Diaporthaceae with special reference to their life histories. Bull. Gov. For. Exp. Stn. No. 226, Tokyo. 242 p.

Raymond, F.L.; Reid, J. 1961. Dieback of balsam fir in Ontario. Can. J. Bot. 39:233-251.

Prepared by D.T. Myren.

3. Red band needle blight
Mycosphaerella pini Rostrup
Plate 3

Hosts: Mainly Austrian pine; rarely Scots pine.

Distribution: Limited distribution; reported from St. John's, Newfoundland, several points in northern and south-central Ontario, and Drummondville, Quebec.

Effects on hosts: Significant defoliation of Austrian pine has been observed in Ontario, but there is no information concerning the effects of this fungus on tree growth. It is conceivable that small trees could be killed by several successive years of severe defoliation. Larger trees would probably experience a reduction in growth rate under similar conditions.

Identifying features: The most characteristic feature of infection by *M. pini* is the reddish bands that encircle the needles. Fruiting structures develop beneath the needle surface in the center of these bands as small, black bodies that enlarge and rupture the needle epidermis. Early infection is characterized by lighter green, yellow, or tan spots on needles. As the infection ages, these spots turn distinctly brown and enlarge to produce characteristic red bands around the needle. Both the brown and reddish discoloration can be seen on green needles, but the reddish areas are most distinct on needles that are dead or recently cast. Infection occurs first on the lower portion of the tree and progresses upwards.

Life history: The fruiting structures of *M. pini* develop in the fall or the spring of the year following infection. Spores are produced in erumpent structures in the reddish bands in late spring, but they can be found throughout the summer. These spores are scattered primarily by rain splash. Although the early symptoms of infection are observed in midsummer, the total period over which infection can occur in eastern Canada is not known. Needles of the current year are not susceptible to infection until midsummer, but second-year and older needles are susceptible throughout the year. Infection could begin in late spring and continue through late summer and fall. Needles with mature fruiting bodies are shed during the summer and fall.

Control: Fungicides provide the most effective control of this disease, but timing of the application is critical. More information is required on the biology of the organism, particularly under the climatic conditions encountered in those areas of eastern Canada where the fungus is found, before a specific time frame for fungicide application can be recommended. Usually an early application to protect the second-year needles and a second application to protect the needles of the current year, which are at first resistant to infection, are required. Some success has been achieved using rigorous sanitation, but this is practical only on a limited number of trees, and removal of infected foliage could ruin the form and appearance of the host.

Additional information: Brown spot needle blight, caused by *Mycosphaerella dearnessii* Barr (see Section 4), is similar to the red band needle blight caused by *M. pini* in many respects but lacks the reddish color in the infected areas on the needles. Ontario is the only province in eastern Canada in which *M. dearnessii* has been found. Collections have been made from Austrian pine, and early symptoms of infection and fruiting are quite similar. The imperfect state of *M. pini*, *Dothistroma septospora* (Dorog.) Morelet, is the only state of the fungus that has thus far been found in eastern Canada. *Mycosphaerella pini* is also known as *Scirrhia pini* Funk & Parker.

Selected bibliography

Funk, A.; Parker, A.K. 1966. *Scirrhia pini* n. sp., the perfect state of *Dothistroma pini* Hulbary. Can. J. Bot. 44:1171-1176.

Hepting, G.H. 1971. Diseases of forest and shade trees of the United States. U.S. Dep. Agric., For. Serv. Agric. Handb. No. 386. 658 p.

Prepared by D.T. Myren.

A

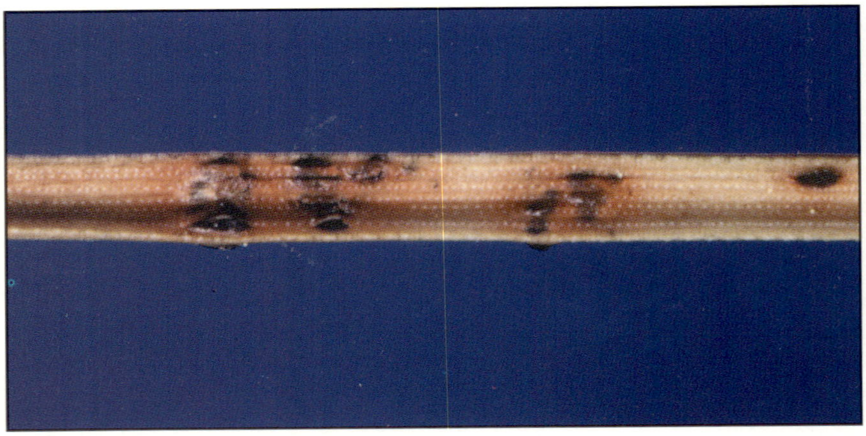

B

Plate 3

A. Austrian pine infected by *Mycosphaerella pini*, the cause of red band needle blight.

B. Fruiting bodies of *Dothistroma septospora*, the imperfect state of the red band needle blight fungus, *Mycosphaerella pini*, and the characteristic red color on an infected Austrian pine needle.

4. Brown spot needle blight
***Mycosphaerella dearnessii* Barr**
Plate 4

Hosts: Mainly Austrian and mugho pine.

Distribution: Known from only four locations in Ontario: Sauble Falls Provincial Park, Inverhuron Provincial Park, Tiny Township (north of Barrie), and Belleville.

Effects on hosts: Brown spot needle blight does not usually kill the host but can cause significant defoliation of the 2- and 3-year-old needles. First-year needles can also be infected when the disease is severe. Portions of the crowns of ornamental mugho pines have been killed by infections of *M. dearnessii.*

Identifying features: Needle symptoms usually appear in the late summer or fall as yellow or brown bands ranging from 1.0 to 3.5 mm in length. The discolored areas are often discrete and frequently resin-soaked. The color of the bands and the adjacent needle tissue may vary somewhat from host to host. Needles die from the tip back, being completely killed by late fall. In the late summer or fall, black fruiting bodies emerge from the discolored spots with a noticeable splitting or tearing of the host epidermis. Some fruiting bodies may not develop until the following spring.

A

B

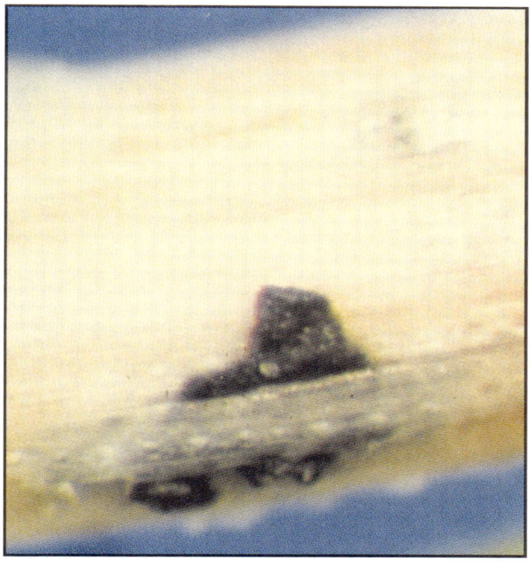
C

Plate 4

A. Mugho pine infected by *Mycosphaerella dearnessii*, the cause of brown spot needle blight.

B. Fruiting structure of *Lecanosticta acicola*, the imperfect state of *Mycosphaerella dearnessii*, on infected mugho pine needles, illustrating the discoloration associated with brown spot needle blight.

C. Fruiting structure of *Lecanosticta acicola*, the imperfect state of *Mycosphaerella dearnessii*, the causal organism of brown spot needle blight, on an infected mugho pine needle.

Life history: Infection is caused by spores exuded from mature fruiting bodies and spread by wind and rain splash. This type of spread usually covers only short distances. Most new infections occur from early to midsummer as the overwintering and newly developed fruiting bodies become active. Some infection also occurs in the fall.

Fruiting bodies begin to form on the needles during the year of infection and may be active late in the fall or the following spring and early summer. Fruiting bodies on needles infected late in the season develop early the following year.

Control: In small areas, control by cutting and destroying infected material has been used with fairly good results, but in most cases fungicides are necessary to control the disease. The Austrian Hills or German varieties of Scots pine, a common host of *M. dearnessii* in many areas, are apparently more resistant to the disease than the Spanish and French-green varieties and should be selected for planting where brown spot needle blight is a problem.

Additional information: The imperfect state of *M. dearnessii* is *Lecanosticta acicola* (Thüm.) H. Sydow and is the only state of the fungus found in Canada. *Mycosphaerella pini* Rostrup (formerly *Scirrhia pini* Funk & Parker) (see Section 3) is similar to *M. dearnessii* (formerly *Scirrhia acicola* (Dearn.) Siggers) but is usually found on Austrian pine in Ontario, and the needles have a characteristic red discoloration in the infected area.

In Ontario, samples of needles for diagnosis of *M. dearnessii* should be collected in early summer, which seems to be the best time to find fresh fruiting bodies.

Selected bibliography

Boyce, J.S. 1961. Forest pathology. 3rd ed. McGraw-Hill Book Co., New York, NY. 572 p.

Punithalingam, E.; Gibson, I.A.S. 1973. *Scirrhia acicola.* Commonw. Mycol. Inst. (CMI) Descr. Pathog. Fungi Bact. No. 367. 2 p.

Prepared by D.T. Myren.

5. Tip blight of conifers
Sphaeropsis sapinea (Fr. : Fr.) Dyko & B. Sutton
Plate 5

Hosts: Mainly Austrian and Scots pine; occasionally lodgepole, mugho, and red pine; rarely Douglas-fir, eastern white and jack pine, and black and Norway spruce.

Distribution: Collected widely in southern and central Ontario and scattered locations in Nova Scotia; reported from Newfoundland and Quebec.

Effects on hosts: Infection by *S. sapinea* can be a serious problem in Christmas tree plantations, windbreaks, and ornamental plantings. The fungus kills the new shoots and can move into older tissue. Repeated attacks disfigure and weaken the host significantly, and younger trees can be killed. In two plantations in southern Ontario, Scots pine 10 m high suffered defoliation levels of 94 and 88% and mortality of 19 and 9%, respectively.

Identifying features: Stunting of shoots and browning of the current year's needles are the most readily observed symptoms. Early stages of infection are often characterized by resin droplets on the new shoots before the needles emerge. Dead shoots from previous years are often present. Small black fruiting bodies may be found on infected tissue, particularly late in the season. Fruiting structures can often be found in the summer by splitting the old shoots and twigs longitudinally. The young fruiting structures can then be seen under the bark. Pitchy cankers may form under the bark of twigs or on the main stem, resulting in the death of branches or entire trees.

Life history: Fruiting bodies of the fungus develop on infected needles and twigs during periods of high rainfall and may remain on infected material for two seasons. Production of spores begins shortly before the buds open and continues throughout much of the growing season. Most infection takes place within the first 2 weeks after bud break. Symptoms begin to appear about 1 month later. Second-year cones are also infected and provide a good overwintering site.

Control: Fungicides provide good control of tip blight. Two applications, one at bud break and one 7–10 days later, provide a reasonable degree of protection. Fungicides do not protect the cones, which remain as a source of infection. Pruning infected material has not been shown to be an effective control measure.

Additional information: *Sphaeropsis sapinea* has been known by several names, *Diplodia pinea* (Desm.) Kickx being the most common. Samples submitted for diagnosis should have fruiting bodies, and it is usually best if older dead shoots are also included. *Sphaeropsis sapinea* has also been implicated in root rot in several areas, including Ontario. Another tip blight, caused by *Kabatina thujae* A. Schneider & v. Arx var. *juniperi* (A. Schneider & v. Arx) Morelet, has been found on ornamental junipers in southern Ontario. The disease causes mortality of the current year's shoots. The fungus fruits in the late spring at the base of the shoots killed the previous year. Rain splash carries the spores to the newly developing shoots, where infection occurs.

Selected bibliography

Hepting, G.H. 1971. Diseases of forest and shade trees of the United States. U.S. Dep. Agric., For. Serv. Agric. Handb. No. 386. 658 p.

Punithalingam, E.; Waterston, J.M. 1970. *Diplodia pinea.* Commonw. Mycol. Inst. (CMI) Descr. Pathog. Fungi Bact. No. 273. 2 p.

Prepared by D.T. Myren.

A

B

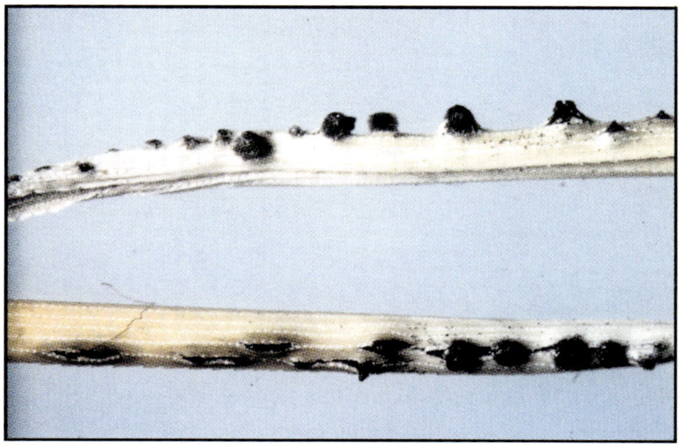

C

Plate 5

A. Tip blight of Austrian pine caused by *Sphaeropsis sapinea.*

B. Buds and needles killed by *Sphaeropsis sapinea,* the cause of tip blight on red pine.

C. Fruiting bodies of *Sphaeropsis sapinea,* a cause of tip blight on infected Austrian pine needles.

6. Shoot blight of aspen
Venturia macularis (Fr. : Fr.) E. Müller & v. Arx
Plate 6

Hosts: Mainly trembling aspen; occasionally hybrid and largetooth aspen; rarely white poplar.

Distribution: Widely distributed throughout aspen stands in eastern Canada.

Effects on hosts: Infection by this fungus causes a shoot blight that results in the cessation of terminal growth and deformation of young aspen. Trees older than 5 years are usually not affected but remain subject to infection when conditions are right. The disease is of little or no economic importance in natural stands, although repeated loss of current growth can result in stag-headed trees and thus delay harvest. Damage to hybrids in plantations can be serious.

Identifying features: The initial symptoms of infection are angular, black leaf spots that usually occur in early summer and often enlarge and coalesce to involve the entire leaf, which then becomes wilted. Apical portions of infected shoots wither and become hook-shaped (resembling a "shepherd's crook") and continue bearing the shriveled, blackened leaves. Later infections occur in early to midsummer on the upper surfaces of older leaves, appearing as discrete brown spots about 0.5 mm in diameter. Under moist conditions, spores are produced in olive-green powdery masses on the infected leaves and shoots.

Life history: The fungus overwinters as both spores and mycelium in infected leaves and shoots. In the spring, the overwintered spores cause the primary infection. Spores produced during the first infection period cause secondary infection on the older leaves. Infection increases rapidly during wet springs but subsides with the onset of warm, dry, summer days.

Control: Fallen, infected leaves should be removed and disposed of in the autumn and early spring. Pruning to remove any overwintering stage on infected shoots also checks the spread of the disease. Fungicides used at bud break have also proven effective as a control measure.

Additional information: The imperfect state of *V. macularis* is *Pollaccia radiosa* (Lib.) Bald. & Cif. *Venturia populina* (Vuill.) Fabric., with the imperfect state *Pollaccia elegans* Servazzi, occurs on balsam poplar and produces a similar disease. Spore sizes and hosts help separate these very similar fungi. The perfect state of *V. macularis* is rarely found.

Selected bibliography

Barr, M.E. 1968. The Venturiaceae in North America. Can. J. Bot. 46:799-864.

Hepting, G.H. 1971. Diseases of forest and shade trees of the United States. U.S. Dep. Agric., For. Serv. Agric. Handb. No. 386. 658 p.

Prepared by D.T. Myren.

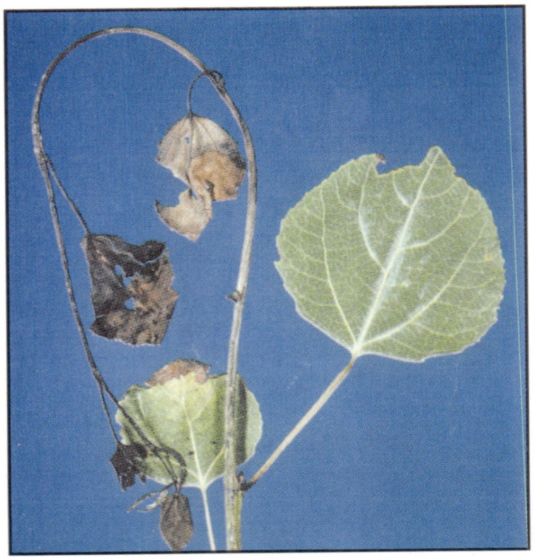

Plate 6

Shoot blight of aspen caused by *Venturia macularis*, with the typical dead leaves and hooking of the infected shoot.

7. Needle blight
Rhizosphaera kalkhoffii Bubák
Plate 7

Hosts: Mainly blue and white spruce; occasionally Norway spruce.

Distribution: Collected from a few locations in central and southern Ontario; known in Quebec.

Effects on hosts: This fungus causes severe defoliation of ornamentals and Christmas trees, reducing both the aesthetic and commercial value of the infected hosts. Small trees can be killed.

Identifying features: Characteristic small, round, black fruiting bodies develop in the stomata on the underside of the needle. They appear as fuzzy black dots, often with white wax from the needle evident on their upper surface. Infection spreads from the bottom of the tree upward and from the base of the branches outward. During the second year of infection, the foliage turns yellow during the summer, becoming brown or purplish brown in the late summer or fall. These needles are often shed at this time. Discoloration is sometimes evident late in the first year of infection.

Life history: Fruiting bodies develop in the stomata in late fall of the year in which infection occurs or early the following spring. Spores from these fruiting bodies are released during wet weather in the late spring or early summer, and rain splash or wind moves them to uninfected needles where new infections can occur.

Control: Fungicides can be used to protect trees in areas where the disease is a problem. However, early detection of the disease is important. As the fungus can be transported on nursery stock, all new planting material should be carefully inspected to make sure it is free of infection. Shearing infected trees should be avoided during wet weather, and infected trees or plantations should be sheared last. Cutting tools should be sterilized with alcohol between cuts when used on infected trees. Blades should be wiped thoroughly with rubbing alcohol between cuts on the same tree and soaked for several minutes between trees.

Additional information: A similar species, *Rhizosphaera pini* (Corda) Maubl., is common in Ontario and is usually found on balsam fir. Both species, *R. kalkhoffii* and *R. pini*, have been recorded on spruce but can be distinguished by spore size.

Selected bibliography

Diamondis, S.; Minter, D.W. 1980. *Rhizosphaera kalkhoffii*. Commonw. Mycol. Inst. (CMI) Descr. Pathog. Fungi Bact. No. 656. 2 p.

Hepting, G.H. 1971. Diseases of forest and shade trees of the United States. U.S. Dep. Agric., For. Serv. Agric. Handb. No. 386. 658 p.

Prepared by D.T. Myren.

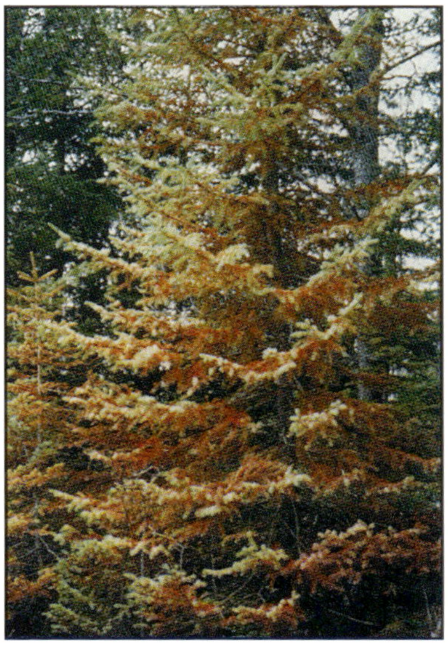

A

Plate 7

A. A black spruce infected by *Rhizosphaera kalkhoffii*, the causal organism of needle blight, on spruce and balsam fir.

B. Fruiting bodies of *Rhizosphaera kalkhoffii*, the causal organism of needle blight, on an infected spruce needle.

B

8. Leaf blister
Taphrina caerulescens (Desm. & Mont.) Tul.
Plate 8

Hosts: Oaks, especially red oak.

Distribution: Found in New Brunswick, Nova Scotia, Ontario, Prince Edward Island, and Quebec.

Effects on hosts: Heavy infections may impair growth of infected trees, particularly if repeated for many years. This fungus is rarely destructive at Canadian latitudes.

Identifying features: The numerous bulging and conspicuous round blisters on leaves, sometimes as many as three or four on the same leaf, occur in early summer. They are at first light green, then grayish, and finally brown. Bulging is not evident at the early stage of disease development but soon becomes quite apparent.

A

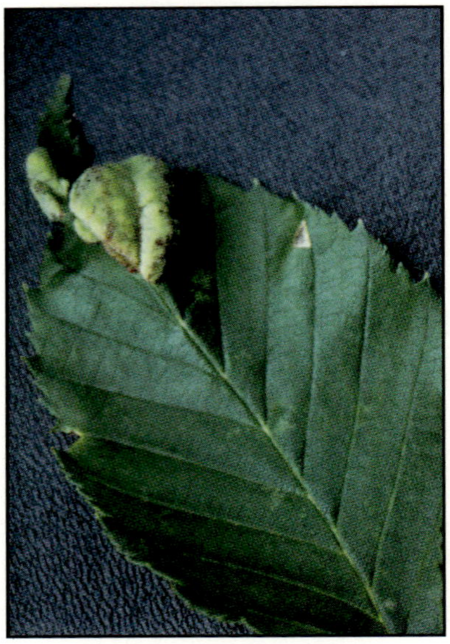

B

C

Plate 8

A. Leaf blister, caused by *Taphrina caerulescens*, on red oak.

B. Leaf blister, caused by *Taphrina carnea*, on yellow birch.

C. Catkin blister, caused by *Taphrina robinsoniana*, on young fruits of alder.

Life history: In midsummer, spores are produced by the fungus between the cuticle and upper epidermis of the leaf. As the spore-producing structures enlarge, they rupture the epidermis and are exposed on the leaf surface. Spores carried by rain and wind overwinter under the bud scales, ready to germinate the following spring and cause new infections. Infection is limited to a short period in the spring, when young, immature leaf tissues are exposed. Cool, wet weather favors the fungus. As the leaves mature, they become more resistant to infection, and the disease ceases to spread.

Control: Fungicide applications are effective but not usually necessary. Trees severely defoliated for several consecutive years should probably be treated. Fungicides can also be used in special cases, when it is desirable to prevent unsightly infections. One application of the chemical 1 or 2 weeks before bud swell usually controls the disease.

Additional information: Other species of *Taphrina* cause similar damage on many other broad-leaved trees. *Taphrina carnea* Johanson occurs mainly on yellow and white birch, *Taphrina dearnessii* Jenkins is found on maple, and *Taphrina robinsoniana* Gies. causes catkin blisters on alder.

Selected bibliography

Boyce, J.S. 1961. Forest pathology. 3rd ed. McGraw-Hill Book Co., New York, NY. 572 p.
Mix, A.J. 1949. A monograph of the genus *Taphrina*. Univ. Kans. Sci. Bull. 33 (1):1-167.

Prepared by R. Cauchon.

9. Leaf blotch
Guignardia aesculi (Peck) V.B. Stewart
Plate 9

Host: Horse-chestnut.

Distribution: Widely distributed throughout eastern Canada.

Effects on host: Horse-chestnut, which is not native to Canada, is widely planted as an ornamental and shade tree. Defoliation due to leaf blotch can be so severe that the value of the tree as an ornamental is significantly reduced. Several successive years of heavy defoliation can weaken the tree, making it more vulnerable to other pest, weather, or site problems. Leaf blotch can be particularly damaging in nurseries where trees can be completely defoliated and growth greatly retarded.

Identifying features: The first indication of the disease is a slight discoloration at the point of infection, which may occur anywhere on the leaf or petiole. The infected areas soon become irregularly shaped, reddish-brown, dead blotches; these enlarge in size until late summer, by which time severely infected leaves have become brittle, brown, and distorted, and some leaves have started to fall. In a fully developed lesion, the center area is dark red to brown with a yellowish margin. Lesions on petioles and fruit appear as elongate, reddish-brown spots. Fruiting bodies of the fungus develop relatively early and appear as small black dots on the recently dead or dying leaf tissue.

Life history: The small black fruiting bodies on the infected leaves produce spores that cause secondary infections throughout the summer. In the late summer and fall, a similar structure is produced, which in turn produces a fruiting body that overwinters in an immature state on the fallen, infected leaves. The overwintering fruiting bodies mature in the spring and forcibly eject spores during wet weather. These spores initiate infection on the newly developing leaves, and the disease cycle is repeated.

Control: A fungicidal spray used when the leaves are half grown can effectively control this disease. Chemical control should be used after a year of severe defoliation and in areas where the disease seems to be an annual problem. Raking and destroying leaves in the fall are fairly effective in checking the spread of the fungus, as they remove the overwintering stage.

Additional information: The imperfect state of *G. aesculi* is *Phyllosticta sphaeropsoidea* Ell. & Ev. Samples for identification should include pressed leaves with visible, tiny black fruiting bodies. The presence of these fruiting bodies and lesions on the leaf stalks and leaflets differentiates leaf blotch from a very common and similar-appearing leaf malady called leaf scorch, which is caused by air pollution (see Section 83) or hot, dry, sunny weather (see Section 90).

Selected bibliography

Boyce, J.S. 1961. Forest pathology. 3rd ed. McGraw-Hill Book Co., New York, NY. 572 p.
Stewart, V.B. 1916. The leaf blotch disease of horse-chestnut. Phytopathology 6:5-19.

Prepared by D.T. Myren.

Plate 9

A. *Guignardia aesculi*, leaf blotch, on horse-chestnut.

B. Leaf blotch, caused by *Guignardia aesculi*, on a horse-chestnut leaf.

A

B

10. Ash leaf rust
Puccinia sparganioides Ell. & Barth.
Plate 10

Hosts: Ash, and cordgrass as the alternate host.

Distribution: Not common in eastern Canada, except in coastal areas of New Brunswick and Nova Scotia, where both hosts occur.

Effects on hosts: On ash, infection usually causes only light discoloration and distortion of leaves and petioles. Distortion and swellings may also occur on twigs. When infection is heavy, affected trees appear scorched by midsummer, and defoliation occurs in late summer. Repeated heavy infection, especially on white ash, results in the death of progressively larger branches, and the entire tree may succumb. The disease is most serious in Nova Scotia, where ornamental trees in many communities are disfigured by repeated attacks.

On cordgrass, leaves turn yellow and then brown.

Identifying features: Swellings on twigs and leaf petioles or circular spots on either side of leaf blades are bright orange in early summer. The color is caused by spores in white, cup-like structures easily visible under a magnifying glass. After the spores are released, the spots fade, but the remnants of the cups remain, pockmarking the affected areas. When infection is heavy, the leaf spots, which coalesce, give the tree an unsightly brownish appearance.

Life history: The majority of rust fungi have complicated life cycles, requiring two unrelated hosts to complete their development, during which they produce several different spore forms. Spores released by the fungus on cordgrass infect ash in the late spring shortly after leaf expansion begins. Another spore type is produced on the infected ash host, and these bright yellow spores infect the blades of cordgrass, the "alternate host." A third spore type, this one capable of infecting cordgrass only, constitutes the "repeating stage" throughout the summer, then in the fall gives way to yet another type of spore. These spores, which are dark in color and thick-walled, are the "overwintering stage." The following spring, they germinate and produce the spores that infect ash, thus completing the cycle.

Control: Sulfur sprays applied shortly after bud break may reduce infection on ash but should be considered only for valuable trees in high-hazard areas.

Selected bibliography

Boyce, J.S. 1961. Forest pathology. 3rd ed. McGraw-Hill Book Co., New York, NY. 572 p.

Ziller, W.G. 1974. The tree rusts of western Canada. Environ. Can., Can. For. Serv., Victoria, B.C. Publ. No. 1329. 272 p.

Prepared by L.P. Magasi.

Plate 10

A. Ash infested by ash rust caused by *Puccinia sparganioides.*

B. Ash leaf with rust (*Puccinia sparganioides*) pustules on the leaf and a rust gall on the petiole.

C. Overmature petiole gall formed in response to infection by the ash rust fungus, *Puccinia sparganioides*.

D. Fruiting of ash rust, *Puccinia sparganioides*, on cordgrass, its alternate host.

A

B

C

D

11. Conifer-aspen rust
Melampsora medusae Thüm.
Plate 11

Hosts: Mainly trembling aspen, eastern cottonwood, and a number of poplar hybrids; alternate hosts are species of Douglas-fir, larches, and jack, red, and Scots pine.

Distribution: Common in Ontario and Quebec throughout the range of aspen and larch; less frequent in the Maritime provinces and Newfoundland.

Effects on hosts: The effect of this rust varies with environmental factors; tree mortality is rare. Hybrid poplars are the most severely attacked. In natural forest, damage to poplar is light, except under extreme epidemic conditions during which young trees may be heavily defoliated. Reduction in diameter growth and delayed development of root system have been recorded as a result of severe infection. Infected trees may also be predisposed to damage by insects and other fungi. During the spring following severe infection, bud opening is delayed and inconsistent; often the flowering fails as well. In some parts of eastern Canada, damage on needles of larches or pines in nurseries can be as important as damage on poplars and cottonwood.

Identifying features: On the undersurface of the poplar leaf, a golden-yellow to orange pustule (1 mm in diameter or less) can be observed in summer months. When favorable conditions for the rust occur, the whole leaf surface is invaded. Late in the season, a brown to black, crust-like structure is formed mainly on the undersurface of the leaf, but it can also appear on the upper leaf surface. The alternate hosts (larch and others) show similar or even smaller spore pustules, isolated or grouped, and yellow-orange in color.

Life history: Spores formed in the fall overwinter on dead poplar leaves and serve to infect the conifer host the following spring. A different type of spore produced on the alternate host spreads the fungus to susceptible poplar leaves, where infection occurs. The spores formed on the poplar leaves intensify and spread the fungus to other poplar leaves during the summer; these spores give rise to the crust-like structure that produces spores that overwinter and start the cycle the following year. Saturated atmosphere and mild temperature are critical factors for new infections. A period of high humidity lasting at

Plate 11

A. *Melampsora medusae*, the cause of conifer-aspen rust, fruiting on hybrid poplar.

B. *Melampsora medusae,* the cause of conifer-aspen rust, fruiting on larch.

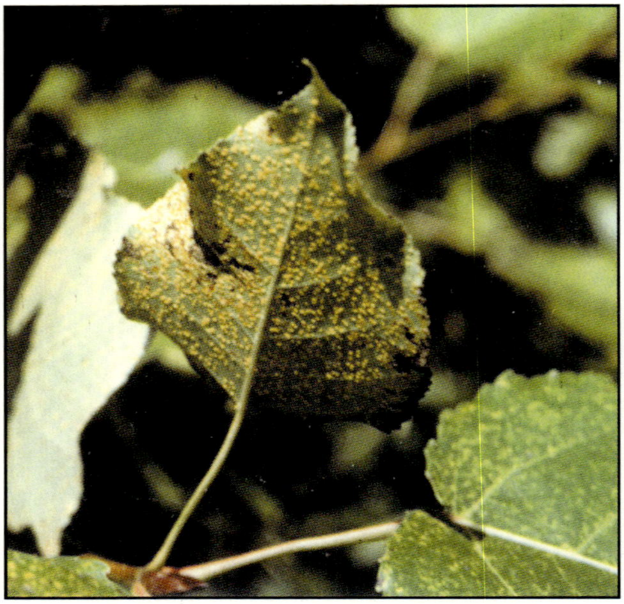

A

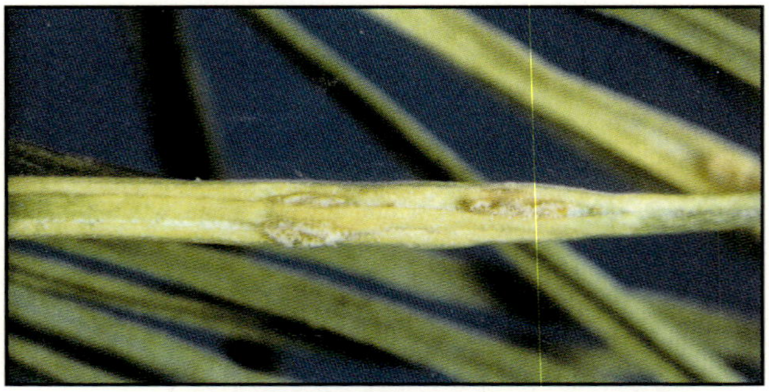

B

least 12 hours is necessary for spore germination and infection to occur.

Control: When choosing a site for aspen or poplar plantations, nearby stands of larches or pines should be avoided, as well as valley bottoms where high humidity is persistent. Resistant clones should be used for establishment of new plantations. In already-established stands, proper aeration of the crown and a balanced potassium soil nutrition should be ensured; excessive nitrogen should be avoided. When years of early infections occur, the foliage of the main or alternate hosts may be sprayed with fungicides. Fungicides are used the first 3 weeks after bud break and might have to be repeated every 3 weeks.

Additional information: The final identification of this rust fungus is based on its microscopic characteristics, because other *Melampsora* species present similar macroscopic features. On willow, several leaf rust fungi produce similar symptoms, with alternate hosts being species of fir, hemlock, larch, and currant. Because it is difficult to differentiate their spore stages on willow, all the North American willow rust fungi have been grouped under *Melampsora epitea* Thüm., also called *M. epitea* complex. Their life cycles follow a pattern similar to the one described for *M. medusae.*

Selected bibliography

Taris, B., ed. 1981. Les maladies des peupliers. Commission Internationale du Peuplier: groupe de Travail des Maladies. Assoc. Forêt-Cellulose, France. 197 p.

Ziller, W.G. 1974. The tree rusts of western Canada. Environ. Can., Can. For. Serv., Victoria, B.C. Publ. No. 1329. 272 p.

Prepared by A. Lavallée.

12. Leaf spot
Stegophora ulmea (Schwein. : Fr.) H. Sydow & Sydow
Plate 12

Hosts: Mainly white elm; occasionally English and Siberian elm and elm hybrids.

Distribution: Found throughout the natural range of elms in eastern Canada.

Effects on hosts: The disease is most damaging during wet seasons; little damage is done under average summer conditions. Severe attacks result in premature defoliation and twig mortality, particularly on recently established trees or in nurseries. Repeated severe attacks weaken the host and predispose it to damage from abiotic factors, insects, and other fungi.

Identifying features: Small whitish or yellowish blotches scattered irregularly on the upper surfaces of the new leaves appear in the spring. Soon after, shiny black imbedded pustules appear in the center of these blotches. These pustules may be surrounded by a "halo." They become grayish with time. Petioles and young twigs may also be infected.

Life history: Spores of the imperfect state, *Asteroma ulmeum* (Miles) B. Sutton, form during the summer in the black pustules on the upper leaf surface and can spread the infection during the growing season. During the winter season, spores of the perfect state (*Stegophora ulmea)* are formed in flask-shaped bodies that protrude from the undersurfaces of the fallen leaves. Spores are discharged from the long necks of these fruiting bodies in the following spring to infect new foliage and begin another infection cycle.

Control: The damage incurred does not usually require specific control measures. However, on ornamentals or in nurseries, the removal and destruction of last year's infected leaves before bud opening significantly reduce the overwintering stage and consequently the number of spores that will be liberated. This measure should reduce initial spring infection. Spray treatments with foliar fungicide can begin when the buds are opening but require strict observation of the concentrations recommended, as new foliar tissues may suffer more damage from incorrectly prepared spray than from the fungus.

Additional information: *Stegophora ulmea* was formerly known as *Gnomonia ulmea* (Schwein.: Fr.)

Thüm. Two other fungi causing leaf spots are frequently recorded on white and English elm. Their imperfect states are *Asteroma inconspicuum* (Cavara) B. Sutton and *Discula umbrinella* (Berk. & Broome) Morelet; they differ slightly from *S. ulmea* in that they appear on both faces of the leaf, and *D. umbrinella* invades midribs and leaf margins.

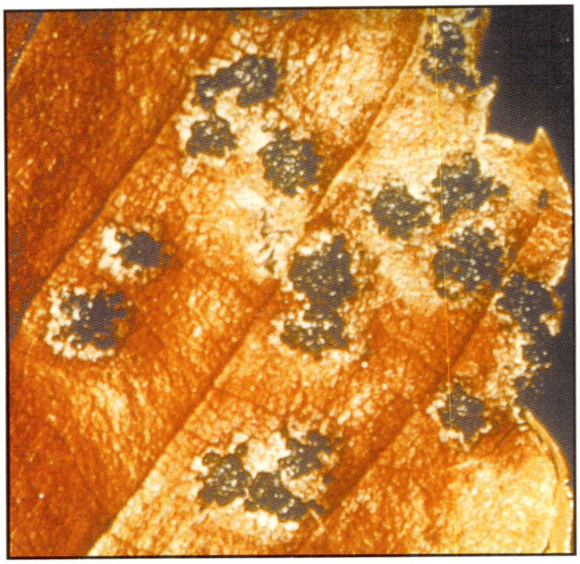

Selected bibliography

Hepting, G.H. 1971. Diseases of forest and shade trees of the United States. U.S. Dep. Agric., For. Serv. Agric. Handb. No. 386. 658 p.

Pomerleau, R. 1938. Recherches sur le *Gnomonia ulmea* (Schwein.) Thüm. (biologie, écologie, cytologie). Contrib. Inst. Bot. Univ. Montreal No. 31. 139 p.

Prepared by A. Lavallée.

Plate 12

Fruiting bodies of *Stegophora ulmea*, a leaf spot fungus, on elm.

13. **Leaf spot**
Septoria aceris **(Lib.) Berk. & Broome**
Plate 13

Hosts: Most species of maple.

Distribution: Widely distributed in eastern Canada.

Effects on hosts: The fungus probably has little effect on the host. Occasionally, when infection is very heavy, the affected leaves are speckled with tiny discolored spots, which may be of concern on ornamental trees.

Identifying features: Small, pinhead-sized spots with a light-colored center and darker border appear on the leaves. One or a few dark fruiting bodies may be visible in the central part of the spots, but only under magnification.

Life history: The fungus is usually visible after the middle of summer. During wet periods, spores ooze out in cream-colored tendrils through a small hole on top of the fruiting body and may cause further infection. The fungus overwinters on fallen infected leaves.

Control: Foliar fungicides would likely reduce infection, although we have no knowledge of any case in which control of this disease was attempted — or deemed necessary.

Additional information: *Septoria aceris* has often been identified and referred to in the literature as *Phloeospora aceris* (Lib.) Sacc.

Selected bibliography

Grove, W.B. 1935. British stem and leaf fungi. Vol. 1. Cambridge University Press, London. 488 p.

Prepared by L.P. Magasi.

Plate 13

Leaf spot on maple caused by *Septoria aceris*. Note the small, dark fruiting bodies of the fungus on several spots.

14. Leaf spot
Phyllosticta minima (Berk. & M.A. Curtis) Underw. & Earle
Plate 14

Hosts: Mainly red, silver, and sugar maple; occasionally other species of maple.

Distribution: Widespread in eastern Canada; can be found wherever maples are present.

Effects on hosts: Leaves spotted with brown blotches lower the aesthetic appearance of infected trees. When infection is heavy, trees appear brown, and premature leaf fall occurs. Several consecutive years of severe infection lower the vigor of affected trees.

Identifying features: Spots are circular or irregular in shape, 5–10 mm in diameter, brown, and surrounded by a purple border. Minute (smaller than a pinhead), black fruiting bodies are scattered over the brown areas. Occasionally, the dried-up central portion of the spot cracks and falls out before the end of the summer.

Life history: Infection occurs during wet periods in the spring when spores, released from the small, black, flask-shaped fruiting bodies on fallen overwintered leaves, penetrate newly developing, expanding leaves. The infected leaf area first becomes lighter green than normal, then turns brown when the infected tissue area dies. The fruiting bodies develop in the summer on the dead areas. Spores are produced inside these flask-shaped fruiting bodies and overwinter to begin the cycle again the following spring.

Control: Raking and destroying (burning or composting) the leaves in the fall reduce infection the following spring. High-value trees can be protected from infection, if necessary, by three applications of a fungicide, at 2-week intervals starting at bud break.

Additional information: The ocellate gall midge, *Acericecis ocellaris* (Osten Sacken), causes a leaf spot somewhat similar to that caused by *P. minima*. However, a depression is almost always recognizable in the center of the spot on the underside of the leaf attacked by the gall midge, and the scattered fruiting bodies are lacking. Saprophytic fungi may develop on the spot, however. *Phyllosticta sorbi* Westend., a closely related fungus, causes similar spots, only smaller in size, on leaflets of American mountain-ash.

Selected bibliography

Boyce, J.S. 1961. Forest pathology. 3rd ed. McGraw-Hill Book Co., New York, NY. 572 p.

Hepting, G.H. 1971. Diseases of forest and shade trees of the United States. U.S. Dep. Agric., For. Serv. Agric. Handb. No. 386. 658 p.

Prepared by L.P. Magasi.

A

B

C

D

Plate 14

A. Leaf spot on maple caused by *Phyllosticta minima.*

B. Fruiting bodies of *Phyllosticta minima* on a leaf spot caused by the fungus on maple.

C. Fruiting bodies of *Phyllosticta sorbi* on a leaf spot caused by the fungus on a mountain-ash leaflet.

D. Leaf spot on maple caused by the insect *Acericecis ocellaris*, which is often mistakenly identified as a leaf spot due to *Phyllosticta minima*.

15. Tar spot
***Rhytisma acerinum* (Pers. : Fr.) Fr.**
Plate 15

Hosts: Mainly red, silver, and sugar maple; occasionally other species of maple.

Distribution: Common throughout the range of its hosts in eastern Canada.

Effects on hosts: The fungus affects parts of the leaf surface, thus reducing the leaf's photosynthetic ability. In the late summer, when the fruiting bodies turn black, the dark blotches on the leaves affect the aesthetic value of the tree. The problem is,

A

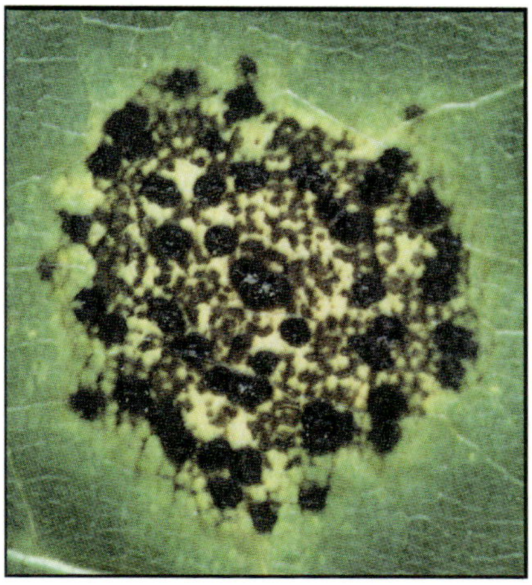

B

C

D

E

Plate 15

A. *Rhytisma acerinum*, the tar spot fungus, infection of a maple leaf.

B. *Melasmia acerina*, the immature state of the tar spot fungus *Rhytisma acerinum*, on maple.

C. *Rhytisma punctatum*, the cause of speckled tar spot, on maple.

D. Tar spots of *Rhytisma punctatum*, the causal organism of speckled tar spot, on an infected maple leaf.

E. Tar spot of willow caused by *Rhytisma salicinum.*

however, rarely serious in ornamental situations. If infection is heavy, premature leaf fall may occur.

Identifying features: One to several black, shiny, raised, tar-like spots up to about 1 cm in diameter occur on the upper surface of the infected leaf in the late summer. Earlier in the summer, these spots appear as light, yellowish-green areas on the leaf that are difficult to identify with certainty as tar spots.

Life history: Spores produced in the black fruiting bodies on fallen leaves are released in the spring and infect newly developed leaves. The fungus affects a small area of the leaf, causing a light, yellowish-green spot. Later in the summer, a thickened fungal structure, the black tar spot, develops. Minute spores develop in this tar spot; their role in spreading the disease is unknown. After the leaves fall in the autumn, another type of spore develops inside the tar spot. These spores mature and are released in the spring.

Control: As the fungus overwinters on fallen leaves, raking and destroying (burning or composting) fallen foliage are recommended as control measures if the condition is serious. Fungicidal treatment starting at bud break is recommended for nursery and other high-value trees.

Additional information: The speckled tar spot fungus, *Rhytisma punctatum* (Pers. : Fr.) Fr., occurs mainly on mountain, silver, and striped maples. Instead of a single large blotch, this fungus produces a group of small, black, pinhead-sized specks, which together form the leaf spot. *Rhytisma punctatum* is much less common than *R. acerinum* in eastern Canada. Another closely related fungus, *Rhytisma salicinum* (Pers. : Fr.) Fr., causes tar spots on willows that are very similar to those caused by *R. acerinum* on maples. The imperfect state of *R. acerinum* is known as *Melasmia acerina* Lév.

Selected bibliography

Boyce, J.S. 1961. Forest pathology. 3rd ed. McGraw-Hill Book Co.,New York, NY. 572 p.

Hepting, G.H. 1971. Diseases of forest and shade trees of the United States. U.S. Dep. Agric., For. Serv. Agric. Handb. No. 386. 658 p.

Prepared by L.P. Magasi.

16. Oak leaf spot
Tubakia dryina (Sacc.) B. Sutton
Plate 16

Host: Red oak.

Distribution: Found in New Brunswick, Ontario, and Quebec.

Effects on host: The fungus can cause considerable leaf spotting and premature defoliation of red oak. Severe defoliation of red oak was noted in late summer in several areas in southwestern Ontario in 1984. There has been no study on the impact of this disease, but it is suspected that several years of severe infection could reduce growth and vigor, particularly of young trees.

Identifying features: Abundant small, circular, brown or reddish-brown spots, 2–5 mm in diameter, are produced on infected leaves. Numerous superficial, grayish to black fruiting bodies are produced in and around the spots on the upper leaf surface. They are usually conspicuous on discolored tissues. On some leaves, black spots are surrounded by a light tan area in which the fruiting bodies develop.

Life history: The life history of this fungus is not completely understood. It probably overwinters on fallen infected leaves in which spores are produced the following spring. The large number of spots noted on some leaves suggests that secondary spread of the pathogen occurs during the summer.

Control: Little information is available on the control of this disease, in part because of the lack of precise information on the life cycle of the fungus. Probably any of the fungicides used to control anthracnose would be satisfactory. Raking and destroying dead leaves would help to control the spread of the fungus.

Additional information: *Tubakia dryina* has a wide host range, although it has been found only on red oak in eastern Canada. The fungus is found in most reference books as *Actinopelte dryina* (Sacc.) Höhnel.

Selected bibliography

Hepting, G.H. 1971. Diseases of forest and shade trees of the United States. U.S. Dep. Agric., For. Serv. Agric. Handb. No. 386. 658 p.

Limber, D.P.; Cash, E.K. 1945. *Actinopelte dryina*. Mycologia 37:129-137.

Prepared by D.T. Myren.

Plate 16

Oak leaf spot caused by *Tubakia dryina*.

17. Ink spot
Ciborinia whetzelii (Seaver) Seaver
Plate 17

Hosts: Mainly trembling aspen; occasionally other poplars and hybrid poplars.

Distribution: Common throughout eastern Canada.

Effects on hosts: Severely infected leaves are completely killed by midsummer or earlier but can persist until autumn. Heavily infected trees may defoliate prematurely. Repeated defoliation can kill young trees and weaken larger ones.

Identifying features: The disease is recognized in early summer by tan to brown areas with concentric white zones on the upper surface of the infected leaf. By midsummer, these spots spread and coalesce, and the entire infected leaf may turn brown. Dark brown to black, raised, hard, oval to circular bodies (sclerotia) develop in the dead areas. These bodies resemble ink spots and are about 2–4 mm in diameter; they are scattered irregularly over the leaf surface. In late summer, the sclerotia fall out, creating "shot holes" in the infected leaves.

Life history: Sclerotia are masses of hyphae that overwinter in leaf debris on the ground. During moist periods in late spring, the sclerotia germinate and produce cup-shaped fruiting bodies. These bodies produce spores, which are discharged forcibly into the air and carried to new, developing leaves on healthy shoots, where infection occurs. The fungus becomes established on the leaves and again produces sclerotia, thus completing the life cycle. No further infection occurs until the following spring.

Control: Removal and burial or burning of fallen infected leaves and sclerotia reduce inoculum levels in plantations and minimize the risk of new infections in the following year. Raking the area beneath the crown may create an unfavorable site for germination of those sclerotia that fell from the leaves.

Additional information: Two other species, *Ciborinia foliicola* (Cash & R.W. Davidson) Whetzel and *Ciborinia candolleana* (Lév.) Whetzel, cause similar diseases in willow and oak, respectively.

Sclerotia of *C. whetzelii* can often be seen as indistinct, somewhat darker brown areas long before true spots develop. Detection of the disease at this stage requires considerable experience, however, as the difference in color from the brown leaf tissue is slight. Samples for diagnosis should contain black sclerotia or else should not be sent before midsummer. Leaves should be pressed properly.

Previously, the pathogen was named *Sclerotinia whetzelii* Seaver and *Sclerotinia bifrons* Whetzel.

Selected bibliography

Boyce, J.S. 1961. Forest pathology. 3rd ed. McGraw-Hill Book Co., New York, NY. 572 p.
Ostry, M.E. 1982. How to identify ink spot of poplars. U.S. Dep. Agric., For. Serv., North Central For. Exp. Stn. Rep. No. HT-53.

Prepared by Pritam Singh.

A

B

C

Plate 17

A. Aspen with infection by *Ciborinia whetzelii*, the causal agent of ink spot.

B. Ink spots caused by infection by *Ciborinia whetzelii*. The spots are sclerotia of the fungus, which will overwinter.

C. *Linospora tetraspora*, the cause of linospora leaf blight on balsam poplar. The black dots are the sclerotia of the fungus.

18. Marssonina leaf spot
Marssonina populi (Lib.) Magnus
Plate 18

Hosts: Poplars.

Distribution: Widespread in eastern Canada.

Effects on hosts: Severe infection may cause premature defoliation of the host. Branch dieback follows repeated defoliation, and the weakened trees are predisposed to other agents. High mortality has been recorded on 1-year-old seedlings of largetooth aspen.

Identifying features: The circular spots, which are very small in the early summer, enlarge up to 2–5 mm in diameter later in the season. They are usually numerous, are orange to chestnut-brown, and frequently have yellow margins. The upper surface is often dotted with tiny grayish to white fruiting structures. Many spots coalesce to form a larger and irregular spot, thus resembling other similar leaf spot diseases; examination under a microscope is necessary to make a proper identification. Although the fungus is usually found on leaf surfaces, it may infect petioles and new shoots.

Life history: Spores produced on the fallen infected leaves of the previous year are dispersed by wind in the spring and are responsible for the development of new infections early in the growing season. The fungus establishes itself in the leaves and, during the summer, produces two types of spores, which are disseminated by rain splash and wind. These spores intensify the infection. Infected leaves often fall prematurely, and the fungus overwinters on this dead material. A new cycle starts the following spring with the production of spores by the fungus on the dead leaves.

Control: Prevention is the best control of this disease, particularly in tree nurseries and plantations. Cuttings for planting should be taken only from healthy shoots, and resistant or the least susceptible clones should be selected for planting. Sanitation — plowing under or removing all plant debris and infected material — also reduces infection.

Additional information: More than seven species of *Marssonina* have been reported on poplars in Canada, and there is considerable confusion in their differentiation. *Marssonina castagnei* (Desm. & Mont.) Magnus, *Marssonina tremulae* (Lib.) Kleb., and *Marssonina balsamiferae* Y. Hirats. have all been reported on poplars in eastern Canada. *Marssonina* species are also found on other tree species: e.g., *Marssonina martini* (Sacc. & Ell.) Magnus on oaks; *Marssonina juglandis* (Lib.) Magnus on hickories, butternut, and black walnut; and *Marssonina betulae* (Lib.) Magnus on birch. Their life histories are similar to the one described.

Selected bibliography
Hepting, G.H. 1971. Diseases of forest and shade trees of the United States. U.S. Dep. Agric., For. Serv. Agric. Handb. No. 386. 658 p.

Taris, B., ed. 1981. Les maladies des peupliers. Commission Internationale du Peuplier: groupe de Travail des Maladies. Assoc. Forêt-Cellulose, France. 197 p.

Prepared by G. Laflamme.

A

B

C

D

E

F

Plate 18

A. Balsam poplar infected by *Marssonina populi*, a causal agent of marssonina leaf spot.

B. Aspen leaves infected by *Marssonina brunnea*, a causal agent of marssonina leaf spot.

C. Fruiting bodies of *Marssonina castagnei*, a causal agent of marssonina leaf spot, on the surface of a white poplar leaf.

D. Fruiting bodies of *Marssonina castagnei*, a causal agent of marssonina leaf spot, on the lower surface of a white poplar leaf.

E. A black walnut leaflet infected by *Marssonina juglandis*, a causal agent of marssonina leaf spot.

F. An oak leaf infected by *Marssonina martini*, a causal agent of marssonina leaf spot.

19. Leaf spot on poplar
Phaeoramularia maculicola (Romell & Sacc.) B. Sutton
Plate 19

Hosts: Mainly trembling aspen; occasionally largetooth aspen.

Distribution: Reported from Newfoundland, Ontario, and Quebec.

Effects on hosts: No detailed information is available concerning the damage done by this disease to its host. The amount of infection on samples, however, indicates that the impact is not severe.

Identifying features: Fruiting bodies appear mainly on the lower leaf surface. The lesions are circular to slightly irregular in shape, approximately 0.5 mm or smaller in diameter, with a green halo and a dark brown, raised border that gets progressively lighter towards the center. In some cases, the whole lesion is brown. The dead tissue on the leaf is evenly pale brown, with the leaf veins medium brown.

Life history: No information is available on the life cycle of *P. maculicola*, but it is suspected that the fungus overwinters in the infected leaves. Infection the following year probably originates from spores liberated from these leaves.

Control: This disease has not been sufficiently damaging to warrant control studies. Therefore, no recommendations can be made.

Additional information: This fungus has often been mentioned in the literature as *Cladosporium subsessile* Ell. & Barth. This disease was selected for coverage because infection by *P. maculicola* is quite similar to secondary infection by *Venturia macularis* (Fr. : Fr.) E. Müller & v. Arx (see Section 6), the fungus that causes shoot blight of aspen. The latter produces its fruiting bodies on the upper leaf surface. These fungi can also be distinguished microscopically.

Selected bibliography
Sutton, B.C. 1970. Forest microfungi. IV. A leaf spot of *Populus* caused by *Cladosporium subsessile*. Can. J. Bot. 48:471-477.

Prepared by D.T. Myren.

Plate 19

Leaf spot on poplar caused by *Phaeoramularia maculicola*.

20. Rhabdocline needle cast
Rhabdocline pseudotsugae H. Sydow
Plate 20

Host: Douglas-fir.

Distribution: Known to occur in a few plantations in New Brunswick, Nova Scotia, and Ontario.

Effects on host: Infected needles discolor and eventually drop off, resulting in a tree with sparse foliage. Heavy repeated infection results in the loss of all but the current year's needles. This causes a loss in growth or, with Christmas trees, lowers the quality, rendering them unmerchantable.

Identifying features: Infected needles have yellow spots in the fall. The following spring, needles turn reddish brown, and heavily infected trees appear scorched. The fungus presses against the outer layers on both surfaces of the needle, creating swellings that split as the fruiting body matures, and the pushed-up pieces of the needle appear as scales.

Life history: Fruiting bodies borne on the 1-year-old needles release spores during rainy periods in

A

B

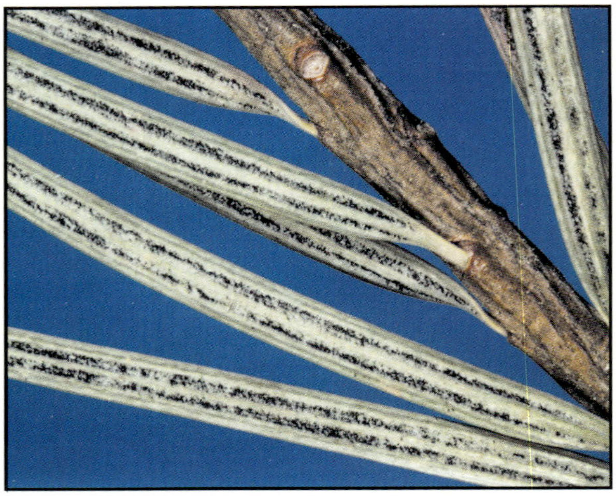

C

Plate 20

A. Severe needle cast of Douglas-fir caused by *Rhabdocline pseudotsugae.*

B. Rupturing of the surface of Douglas-fir needles by the developing fruiting bodies of the needle cast fungus, *Rhabdocline pseudotsugae.*

C. Swiss needle cast, caused by *Phaeocryptopus gaeumannii*, on Douglas-fir needles.

the spring. These spores infect the developing current year's needles. New fruiting bodies mature under the surface of the needle the following spring, by which time the needle changes color. Infected needles are cast shortly after the spores are released.

Control: Fungicidal application at the time of spore release is known to provide adequate protection, but it is practical only in nurseries and high-value plantations. Removal of severely infected trees from plantations also reduces infections.

Additional information: Resistance to infection varies among varieties of Douglas-fir. Green (or Pacific Coast) Douglas-fir is more resistant than either gray (or intermountain) or blue (or Rocky Mountain) Douglas-fir. Christmas tree growers are advised to take this into consideration when buying seedlings. *Phaeocryptopus gaeumannii* (Rohde) Petrak, the Swiss needle cast fungus, has been reported with *R. pseudotsugae* in New Brunswick, Nova Scotia, and Ontario. This fungus produces small, rounded, black fruiting bodies on the undersurface of the needle and is also found only on Douglas-fir.

Selected bibliography

Boyce, J.S. 1961. Forest pathology. 3rd ed. McGraw-Hill Book Co., New York, NY. 572 p.

Parker, A.K.; Reid, J. 1969. The genus *Rhabdocline*. Can. J. Bot. 47:1533-1545.

Prepared by L.P. Magasi.

21. Needle cast
Isthmiella faullii (Darker) Darker
Plate 21

Host: Balsam fir.

Distribution: Found throughout the range of its host in eastern Canada.

Effects on host: This needle cast is the most common and most destructive of the needle casts on balsam fir, but it does not pose a serious problem on older trees. It can cause severe defoliation on seedlings, resulting in reduced growth or even mortality.

Identifying features: Infected needles start to change color in the spring of their second growing season and are totally brown by midsummer. The first fruiting bodies of the fungus are formed on the upper surfaces of these needles and are nearly the same color. They are slightly raised, usually in a double row (sometimes in a single row), and run the full length of the needle in a sinuous or labyrinthine pattern. These fruiting bodies are a very important identifying feature but can be seen only through a hand lens. A second type of fruiting structure forms on the lower surfaces of infected needles in their third growing season. These fruiting bodies are usually evident by midsummer and appear as a single black line running the length of the needle.

Life history: Spores are discharged from the fruiting structures on the 3-year-old needles during the midsummer and cause infection of the first-year needles. The following year, the first type of fruiting body develops on the upper surface of the infected needle; these fruiting bodies discharge their spores later in the summer or early fall. The function of these spores is not understood, but they are suspected to act in the fertilization of other fungal structures on the needle, which give rise to the second type of fruiting body. These second fruiting bodies mature during the midsummer on the needles in their third year and liberate the spores that initiate the new infections.

Control: Control is generally not undertaken unless the trees have considerable value, such as nursery stock, Christmas tree plantations, and ornamentals. Removal of infected foliage provides some measure of control. A fungicide applied at the time the spores are being liberated provides adequate protection. In both cases, timing is important.

Additional information: In early literature, *I. faullii* was known as *Bifusella faullii* Darker. A similar fungus, *Isthmiella crepidiformis* (Darker) Darker, is common on black spruce. In the Maritimes, another needle cast fungus, *Lirula nervata* (Darker) Darker (see Section 22), causes significant damage to balsam fir. Samples submitted for identification of *I. faullii* should include both the 2- and 3-year-old needles.

Selected bibliography

Boyce, J.S. 1961. Forest pathology. 3rd ed. McGraw-Hill Book Co., New York, NY. 572 p.

Darker, G.D. 1932. The Hypodermataceae of conifers. Arnold Arboretum Contrib. No. 1. 131 p.

Darker, G.D. 1967. A revision of the genera of the Hypodermataceae. Can. J. Bot. 45:1399-1444.

Prepared by D.T. Myren.

Plate 21

Black linear fruiting bodies of the needle cast fungus, *Isthmiella faullii*, on the undersides of balsam fir needles at the end of their second year.

22. Needle cast *Lirula nervata* (Darker) Darker

Plate 22

Host: Exclusively balsam fir.

Distribution: Widely distributed throughout eastern Canada.

Effects on host: Infection results in loss of needles; heavy infection degrades Christmas trees and causes growth reduction.

Identifying features: Infected needles are brown. The fruiting body appears as a black line running the entire length of the needle along the underside of the midrib, with a narrow slit that opens up when the needle is wet, exposing a milky, shiny surface. A thinner, more or less continuous, superficial line appears in the middle of the upper side of the needle. Infected needles may fall after spores are released but often persist for the rest of the year.

Life history: The spores produced in the fruiting body the previous year mature in the late spring or early summer. The lips of the fruiting body open in wet weather, and the spores are discharged and carried to new needles by wind or rain drops. The spore germinates and enters the needle, and new infection results. Infected needles turn yellowish, then brown. The fruiting bodies develop in the late summer of the first year after infection and release the spores in the spring of the second year.

Control: Fungicides applied at the time of spore discharge provide adequate protection, but spraying should be considered only for high-value plantations or ornamental trees, and only if fruiting bodies are abundant from previous infection.

Additional information: A similar fungus, *Lirula mirabilis* (Darker) Darker, differs mainly in that there are two raised lines the color of the needle, one along each edge on the upper surface of the infected needle, instead of the central black line as with *L. nervata*. In early literature, these species were called *Hypodermella mirabilis* Darker and *Hypodermella nervata* Darker. These fungi are also similar to another needle cast fungus, *Isthmiella faullii* (Darker) Darker (see Section 21). Trees infected by *I. faullii* are generally younger than those infected by *L. nervata* and *L. mirabilis*; as well, the color of infected needles is lighter, and the configuration of the fungal structures on the upper side of the needles is different. (There are also major microscopic differences.)

Selected bibliography

Boyce, J.S. 1961. Forest pathology. 3rd ed. McGraw-Hill Book Co., New York, NY. 572 p.

Darker, G.D. 1932. The Hypodermataceae of conifers. Arnold Arboretum Contrib. No. 1. 131 p.

Hepting, G.H. 1971. Diseases of forest and shade trees of the United States. U.S. Dep. Agric., For. Serv. Agric. Handb. No. 386. 658 p.

Prepared by L.P. Magasi.

A

B

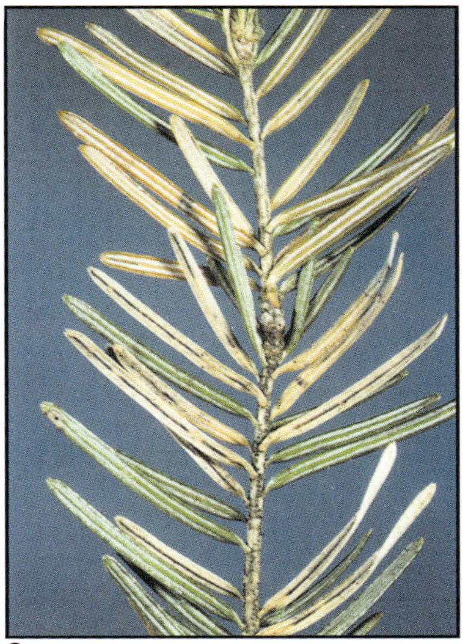

C

Plate 22

A. Needle cast caused by *Lirula nervata* on a balsam fir.

B. Needle cast caused by *Lirula nervata* on balsam fir, showing mortality of needles at the end of their second year.

C. Black linear fruiting bodies of the needle cast fungus, *Lirula nervata*, on the undersurfaces of balsam fir needles.

23. Larch needle cast
***Hypodermella laricis* Tubeuf**
Plate 23

Host: Tamarack.

Distribution: Widespread, but occurs only in patches throughout the range of its host in eastern Canada.

Effects on host: The fungus is not known to cause significant mortality in natural stands and is not currently considered a serious forest pest. It kills the needles and many of the short secondary shoots. Repeated infections cause death of spur shoots. Loss in growth is the most serious damage resulting from severe infections, but trees regain vigor and recover during periods of low infection.

Identifying features: The infected needles turn yellow, then yellow-orange, thus giving a scorched appearance to severely infected trees in the early summer. In midsummer, wilting of the shoot takes place, and the dead needles turn reddish brown. This is followed by the appearance of oblong to elliptical, black fruiting bodies on the upper surface of the needle. Infected and dead needles are retained on the tree for up to 2 years.

Life history: The fungus infects larch within 2 weeks of needle elongation during wet weather in the late spring or early summer. Fruiting bodies on the upper surfaces of dead needles produce minute spores the following spring and are dispersed by rain. These are suspected to function in fertilization. Later in the season, a second type of spore is produced in more prominent, dull black, oblong to elliptical fruiting bodies on the upper surfaces of needles. These second spores overwinter on the dead needles and are released the following spring, coincident with bud break on larch. They are dispersed by wind and initiate new infections on healthy trees.

Control: Because the disease is not considered economically important, no controls have been attempted in forests. However, regular application of a fungicide throughout the growing season is recommended for ornamental and nursery trees. Some control can also be obtained by transplanting stock at the end of the first year, keeping transplant beds in a distant part of the nursery, or cultivating the old beds to bury infected needles, before replanting.

Selected bibliography
Boyce, J.S. 1961. Forest pathology. 3rd ed. McGraw-Hill Book Co., New York, NY. 572 p.
Garbutt, R.W. 1984. Foliage diseases of western larch in British Columbia. Environ. Can., Can. For. Serv., Pac. For. Res. Cent. Pest Leaflet No. 71. 4 p.

Prepared by Pritam Singh.

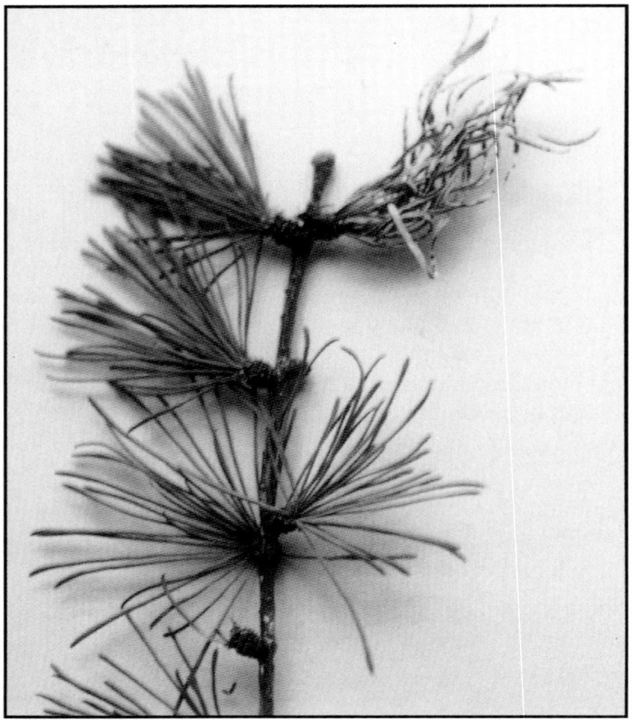

Plate 23

Needle damaged by *Hypodermella laricis*, the causal agent of larch needle cast, on tamarack.

24. Needle cast
Cyclaneusma minus (Butin) DiCosmo, Peredo & Minter
Plate 24

Hosts: Mainly Scots pine; occasionally mugho pine.

Distribution: Reported from New Brunswick, Ontario, and Quebec.

Effects on hosts: Infection by *C. minus* can cause significant damage to ornamentals and Christmas trees. Severe needle cast reduces both the aesthetic and commercial value of the host but is usually not fatal. Several years of severe infection can result in reduced growth and loss of vigor, thus predisposing infected trees to attack by other pathogens or insects.

Identifying features: The first easily recognized symptoms occur in the late summer or fall when the needles in their second and third years begin to turn yellow and brown bars develop across their lengths. Later in the fall, fruiting bodies form on the bars and eventually over the entire needle. The fruiting bodies are about 0.5 mm in length and can be seen by the unaided eye. The epidermis of the needle splits longitudinally, and the margins fold back to the sides, exposing the whitish spore layer of the fruiting body. When closed, the fruiting bodies appear tan, and the split may be evident under a hand lens. Most yellowed needles are cast in the fall and winter.

Life history: The spores of *C. minus* are discharged throughout the growing season, but most infections occur over a 1- or 2-week period in the spring or early summer. These infections are on second- or third-year needles. Needles of the current year can also be infected but are susceptible only after needle elongation has been completed. There is a short period in the mid- to late summer when infection of new needles can occur. Older needles may also be infected at this time. The spores causing infection during spring or early summer and mid- to late summer are

A

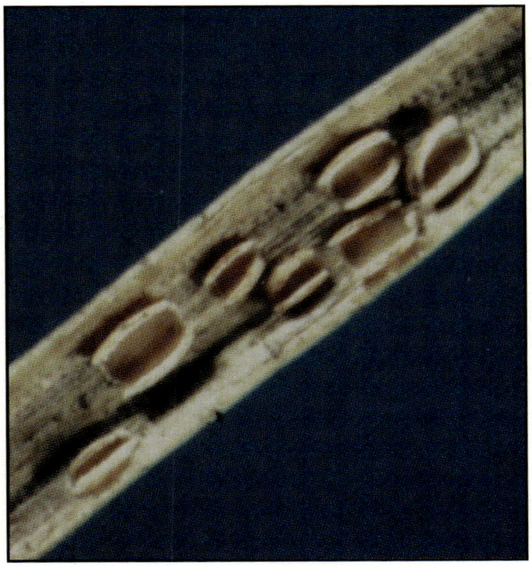

B

Plate 24

A. A Scots pine infected by the needle cast fungus, *Cyclaneusma minus.*

B. Open fruiting bodies of *Cyclaneusma minus* on a Scots pine needle.

produced from fruiting bodies that develop on needles cast during the previous fall and winter. A third infection period occurs in the fall and is believed to be caused by spores from fruiting bodies formed in the late summer on third-year needles. These needles are cast in the early fall. A fourth infection period occurs in the late fall, with spores produced from fruiting bodies on infected needles that are just being cast or will be cast during the winter.

Once infection occurs, a period of 10–15 months may elapse before the onset of symptoms. For example, needles infected in 1980 showed symptoms in 1981 and were cast in the fall and winter of 1981–1982. These needles bore the fruiting bodies that caused the late fall infection in 1981 and the spring and summer infections of 1982. The incubation period can be shorter under some circumstances, and the months during which infections occur vary from year to year and between geographic areas. Also, depending on weather conditions, some of the infection periods may not develop every year.

Control: Fungicides can be used to control this fungus, but a complete spray schedule has not been developed for eastern Canada. Variations in weather conditions from year to year affect the number of fungicide applications required. Sanitation has been tried on a limited scale, but the results have not been encouraging.

Additional information: A similar fungus, *Cyclaneusma niveum* (Pers. : Fr.) DiCosmo, Peredo and Minter, can easily be mistaken for *C. minus*, but it is usually saprophytic. A laboratory examination is needed to differentiate these fungi. Collections of needles with fresh fruiting bodies provide the best samples, but needles with symptoms only might be adequate.

For many years, the fungus was called *Naemacyclus niveus* (Pers. : Fr.) Fuckel ex Sacc., but recent studies revealed the presence of a second species, *N. minor*, and the two species were subsequently placed into their present genus.

Selected bibliography

Merril, W.; Kistler, B.R. 1980. Infection periods in naemacyclus needlecast of Scots pine. Plant Dis. 65:759-762.

Millar, C.S.; Minter, D.W. 1980. *Naemacyclus minor.* Commonw. Mycol. Inst. (CMI) Descr. Pathog. Fungi Bact. No. 659. 2 p.

Prepared by D.T. Myren.

25. Tar spot needle cast
Davisomycella ampla (J. Davis) Darker
Plate 25

Host: Jack pine.

Distribution: Collected in New Brunswick, Nova Scotia, Ontario, and Quebec.

Effects on host: Heavy infection can result in the loss of all but the current year's needles. Some reduction in growth and vigor is expected if severe infection continues for several successive years.

Identifying features: The presence of elliptical, raised, black fruiting structures in buff areas on 1-year-old needles is the best field characteristic and is usually evident in late spring to early summer. The buff areas are often bordered by a brown zone and may become evident in late summer on the needles of the current year. Severe infection can cause loss of all but the current year's needles, giving the tree a sparse appearance.

Life history: The spores mature in late May or June on the needles infected the previous year. The fruiting body splits open in wet weather, allowing spores to be dispersed by rain splash and wind. Spores landing on the current year's needles will initiate the new infection. The infected 1-year-old needles are shed at about the time of spore discharge.

Control: Fungicides applied during the period of spore discharge provide adequate protection. Control is advisable only for high-value plantations and ornamental trees.

Additional information: In early literature, *D. ampla* was referred to as *Hypodermella ampla* (J. Davis) Dearn. A similar fungus, *Davisomycella fragilis* Darker, also occurs on jack pine. Its fruiting body is usually found on totally brown needles. Samples of *D. ampla* should be collected during the early summer, when mature spores are present.

Selected bibliography

Hepting, G.H. 1971. Diseases of forest and shade trees of the United States. U.S. Dep. Agric., For. Serv. Agric. Handb. No. 386. 658 p.

Minter, D.W.; Gibson, I.A.S. 1978. *Davisomycella ampla*. Commonw. Mycol. Inst. (CMI) Descr. Pathog. Fungi Bact. No. 561. 2 p.

Prepared by D.T. Myren.

Plate 25

A. Tar spot needle cast of jack pine caused by *Davisomycella ampla*. The foliage of the current year has just reached full development, and the dead 1-year-old needles will soon be cast.

B. Fruiting bodies of the tar spot needle cast fungus, *Davisomycella ampla*, on jack pine needles in the early summer of their second year.

A

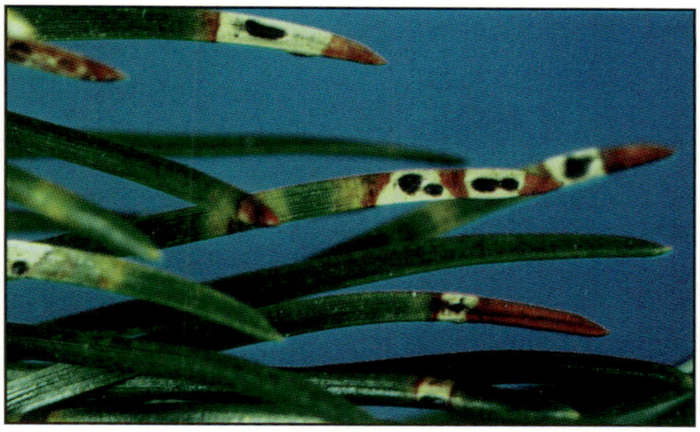

B

26. Needle cast of pine
Lophodermium seditiosum Minter, Staley & Millar
Plate 26

Hosts: Austrian, red, and Scots pines.

Distribution: Common throughout eastern Canada.

Effects on hosts: Infection causes premature fall of needles and may result in the death of seedlings and sapling-sized trees. The disease also affects growth, resulting in stunted seedlings or trees. The damage is usually more severe and conspicuous on the lower branches, but the fungus can infect the whole tree. Generally the trees survive but become unmerchantable as ornamentals or Christmas trees. Seedlings in nurseries may be killed, and damage can be so extensive that entire beds are destroyed.

Identifying features: The first conspicuous symptoms of the disease are small brown spots that develop on the needles in the spring. As the spots enlarge, the needles begin to yellow, then turn brown, and die by early summer. Small black fruiting bodies that develop on the dead needles are quite flat and may be somewhat elongated. Later in the summer, elliptical, grayish to black fruiting bodies are noticeable on all sides of the brown needles. In moist weather, these football-shaped, shiny black fruiting bodies break through the needle epidermis and develop a slit down the middle from which spores are liberated.

Life history: Infection occurs in late summer on the current year's needles but does not become evident until the following spring. The needles die in early summer, and the fruiting bodies of the imperfect state of the fungus develop on the dead tissue. Spores produced by this state do not cause infection but are suspected of being involved in fertilization, giving rise to the perfect state. Later in the season, the fruiting bodies of the perfect state develop and liberate their spores in wet weather. These spores are dispersed by wind and rain splash and cause new infections on the current year's needles. Fruiting can occur on dead needles still attached to the tree or on fallen needles.

Control: Needle casts rarely cause significant damage in forests but can be serious in Christmas tree plantations, windbreaks, and ornamentals. Fungicides provide good control in these situations. The disease can be so serious on seedlings that many forestry nurseries include fungicide application to control needle cast as part of their regular operation. Irrigation of problem areas in nurseries should be done in the morning. Sanitation is also recommended. Infected nursery stock, even though surviving, should not be used in planting because of the possibility of introducing the fungus into new areas.

Additional information: For many years, *Lophodermium pinastri* (Schrader : Fr.) Chev. was considered to be the cause of *Lophodermium* needle cast on pine in North America. Recent research has shown this fungus to be saprophytic, and *L. seditiosum* is now recognized as the only pathogenic *Lophodermium* on pine in eastern Canada. Most early studies reporting damage by *L. pinastri* were probably dealing with *L. seditiosum.* Very sharp black zone lines across the needles help distinguish *L. pinastri* in the field. *Lophodermium seditiosum* may have a few diffuse brown zone lines but no black lines. There are other species of *Lophodermium* known on pine in eastern Canada, and an examination under the microscope is often needed to make an identification. Samples sent for identification should include mature fruiting bodies on the 1-year-old needles.

Selected bibliography

Minter, D.W. 1981. *Lophodermium* on pines. Commonw. Mycol. Inst., Kew, Surrey, England. Mycol. Pap. No. 147. 54 p.

Nichols, T.H.; Skilling, D.D. 1974. Control of *Lophodermium* needlecast disease in nurseries and plantations. U.S. Dep. Agric., For. Serv., North Central For. Exp. Stn., St. Paul, MN. NC-Res. Pap. 110. 11 p.

Prepared by D.T. Myren and Pritam Singh.

A

B

Plate 26

A. A red pine tree infected by *Lophodermium* needle cast fungus.

B. Fruiting bodies typical of *Lophodermium* needle cast fungus on an infected pine needle.

27. Fir broom rust
Melampsorella caryophyllacearum Schröter
Plate 27

Hosts: Balsam fir, with chickweed as the alternate host.

Distribution: Found in scattered patches throughout the range of its hosts in eastern Canada.

Effects on hosts: The disease is usually not severe in balsam fir, because the infection is primarily on branches, and the trunk is not often involved. The damage can result in abnormal shoot growth, malformation and swelling of shoots, tree dieback, spiked tree tops, bole deformation, and branch mortality. Tree mortality can result, but it is rare. The brooms and their cankers can also serve as infection courts for decay fungi, thereby increasing cull. Cull due to trunk infection is a major cause of economic loss caused by the fungus. Trees may also be culled if brooms become large. The disease also reduces growth in height and diameter, resulting in volume loss.

On chickweed, the rust fungus causes leaf and shoot blight.

Identifying features: On balsam fir, the most conspicuous symptoms of the disease are yellowish-green, upright, dense, broom-like structures that develop in the early part of summer. These are produced by the abnormal growth of large numbers of upright lateral shoots from the infected branch or an infected area. The needles on these brooms are dwarfed and yellowish in color. Yellowish-orange spores are produced in pustules on these needles during the summer. Infected needles drop off at the end of each growing season. As well, spindle-shaped swellings or galls often develop on the branches and trunks of infected trees.

On chickweed, the rust is seen as small, orange-red pustules or as whitish to pale reddish spots on the undersurfaces of leaves.

A

B

Plate 27

A. Witches'-broom of fir broom rust on balsam fir caused by *Melampsorella caryophyllacearum*. Note the upright form of the broom and the chlorosis of the needles.

B. Distorted balsam fir needles from a witches'-broom of fir broom rust caused by *Melampsorella caryophyllacearum*. Note the small white pustules, which will soon burst open to release the spores that infect the alternate host.

Life history: The fungus spends its life on two different host species, balsam fir and chickweed. In the spring or early summer, spores develop in two rows of small, round, orange-yellow blisters on both surfaces of yellow, dwarfed balsam fir needles on a broom. These spores function in fertilization. By midsummer, another type of fruiting body develops in two rows on the undersurfaces of needles and produces a second type of spore. These spores are dispersed by wind and infect the foliage of the alternate host (chickweed). After a few weeks, a third spore type is produced on chickweed, which carries the disease again to chickweed and results in an intensification of infection on this host. At the end of the summer or in early fall, white to pale reddish spots appear on the chickweed, which produce a fourth type of spore on chickweed leaves. These spores overwinter, germinate the following spring, and release another type of spore, which infects balsam fir. The infected needles drop each year, but the disease persists in the woody tissue of the broom so that the new shoots growing from the broom and needles become infected as soon as they begin growth each year.

Control: Pruning and burning of infected branches and removal of trees with stem galls are recommended to reduce the incidence and spread of the disease. Elimination of chickweed, the alternate host, in the vicinity of balsam fir also effectively controls the spread of the disease, but it is generally not practical. Large-scale control of the rust in forests is not considered economical.

Selected bibliography

Boyce, J.S. 1961. Forest pathology. 3rd ed. McGraw-Hill Book Co., New York, NY. 572 p.

Singh, P. 1978. Broom rust of balsam fir and black spruce in Newfoundland. Eur. J. For. Pathol. 8:25-36.

Ziller, W.G. 1974. The tree rusts of western Canada. Environ. Can., Can. For. Serv., Victoria, B.C. Publ. No. 1329. 272 p.

Prepared by Pritam Singh.

28. Witches'-broom of blueberry
Pucciniastrum goeppertianum (Kühn) Kleb.
Plate 28

Hosts: Balsam fir, with blueberry as the alternate host.

Distribution: Occurs throughout eastern Canada.

Effects on hosts: On balsam fir, the infected needles shrivel up and drop prematurely. When infection is heavy, the loss of foliage may result in growth loss, particularly on young trees. With Christmas trees, infection may result in a reduction in grade and consequently significant economic loss. On blueberry, infected shoots do not produce fruit, so there is a loss in production.

Identifying features: On balsam fir, small, orange-yellow blisters develop in early summer on the undersides of current-year needles on both sides of the main vein. The infected needles turn yellow, then brown, and finally dry up and fall. On blueberries, clumps of swollen branches occur, forming witches'-brooms. Older swollen shoots are dry and cracked.

Life history: This fungus overwinters as a resting spore in the bark of infected blueberry shoots. The

Plate 28

A. Balsam fir trees infected by *Pucciniastrum goeppertianum*, the cause of witches'-broom of blueberry.

B. Fruiting of *Pucciniastrum goeppertianum*, the cause of witches'-broom of blueberry, on balsam fir needles.

C. Witches'-broom of blueberry caused by *Pucciniastrum goeppertianum.*

D. *Pucciniastrum epilobii,* the fireweed rust fungus, on balsam fir trees.

E. Fruiting of *Pucciniastrum epilobii,* the fireweed rust fungus, on a balsam fir needle.

F. Fireweed, the alternate host of *Pucciniastrum epilobii.*

G. *Uredinopsis* sp. on balsam fir. Note the spore pustules containing the white spores. The members of this genus alternate on ferns.

A

B

C

D

E

F

G

bark is sloughed off in the spring, and the resting spores produce the spores that infect the new needles of balsam fir. In about 2 weeks' time, another type of spore develops in small white sacks on the undersides of the needles. These spores are orange-yellow in color and infect only the newly expanded blueberry shoots. The year following infection, a witches'-broom is formed as a result of a proliferation of swollen branches. Witches'-brooms are perennial.

Control: When both hosts have commercial value, a choice is necessary. One of the hosts must be eliminated from the area, or chemical control can be used. Herbicides to destroy broomed blueberry or fungicides applied to balsam fir just after bud break are probably adequate.

Additional information: A similar needle rust fungus, *Pucciniastrum epilobii* Otth, alternates between balsam fir and fireweed (*Epilobium sp.*). The life cycle of the fireweed rust fungus differs from that of *P. goeppertianum* in that it has a spore type on the alternate host that infects other alternate host plants. In some open areas where fireweed abounds, damage to balsam fir by this rust exceeds that caused by the blueberry rust. Infection by *P. epilobii* occurs somewhat earlier in the summer, and the same tree may be infected by both fungi. Elimination of fireweed in and around Christmas tree areas is recommended as a control method for *P. epilobii.*

Other needle rust fungi, which produce white instead of orange-yellow spores on balsam fir, also occur in eastern Canada. These are species of *Uredinopsis* and *Milesia*, with various ferns serving as alternate species.

Selected bibliography

Boyce, J.S. 1961. Forest pathology. 3rd ed. McGraw-Hill Book Co., New York, NY. 572 p.

Hepting, G.H. 1971. Diseases of forest and shade trees of the United States. U.S. Dep. Agric., For. Serv. Agric. Handb. No. 386. 658 p.

Ziller, W.G. 1974. The rusts of western Canada. Environ. Can., Can. For. Serv., Victoria, B.C. Publ. No. 1329. 272 p.

Prepared by L.P. Magasi.

29. Mountain-ash–juniper rust
Gymnosporangium cornutum Arthur ex Kern
Plate 29

Hosts: Mainly common juniper, with species of mountain-ash as the alternate host.

Distribution: Sporadic in small patches throughout eastern Canada wherever the hosts occur.

Effects on hosts: On juniper, the rust produces slight fusiform swellings on twigs and branches and may cause mortality. It induces faint yellowing or chlorosis of entire needles. On mountain-ash, it causes a leaf spot and may result in premature defoliation.

Identifying features: On juniper, the needles become chlorotic and develop pulvinate, chocolate-brown, pustule-like fruiting bodies. The infected twigs and branches exhibit slight but conspicuous swellings.

The most obvious symptoms on mountain-ash are reddish-brown or purple leaf spots in early summer and the presence of cylindrical, slightly curved, horn-like structures, in groups, on the undersurface of the infected leaf, usually in late summer.

Life history: The fungus fruits on the juniper host in the spring, producing dark brown, cushion-shaped fruiting bodies on the needles or swollen areas on the twigs. The spores produced germinate in place and produce the spores that cause infection on the mountain-ash. Infection of this host results in pale yellow spots on the upper leaf surface in which small yellow pustules develop. These pustules turn black as they mature and produce spermatia, which are involved in fertilization. The next spore stage develops on the lower leaf surface and directly below the pustules on the upper surface. Brownish, horn-like structures are formed, and brown spores are produced within these structures, which tear open in late summer to early fall and release the spores. These spores cause infection of the juniper host.

Control: Because juniper and mountain-ash are of little economic importance in forests, and because the rusts occur in small patches, no controls are recommended. However, infected branches on ornamental juniper trees are often pruned to reduce infection.

Additional information: The spore stages on juniper are very inconspicuous but could probably be found in the spring on juniper located near mountain-ash infected the previous summer. Most collections of this rust are from mountain-ash.

Selected bibliography

Ziller, W.G. 1974. The tree rusts of western Canada. Environ. Can., Can. For. Serv., Victoria, B.C. Publ. No. 1329. 272 p.

Prepared by Pritam Singh.

Plate 29

A. Fruiting of *Gymnosporangium cornutum*, the mountain-ash–juniper rust fungus, on the juniper host. The gelatinous spore horns develop only in wet weather. (Photograph courtesy of M. Dumas.)

B. Fruiting of *Gymnosporangium cornutum*, the mountain-ash–juniper rust fungus, on mountain-ash leaflets. (Photograph courtesy of M. Dumas.)

A

B

30. Cedar-apple rust
Gymnosporangium juniperi-virginianae Schwein.
Plate 30

Hosts: Mainly eastern red cedar; alternate hosts are species of apple, including crab apple.

Distribution: Common in southern Ontario and southern Quebec.

Effects on hosts: The rust affects leaves, branches, and fruits of apple and crab apple trees, causing premature defoliation and dwarfing of fruit. Severely affected or highly susceptible apple or crab apple trees may be killed.

The rust fungus infects and kills leaves and small branches of eastern red cedar and produces unsightly galls. The fungus girdles and kills the smaller twigs. When many galls are produced on a tree, its growth is reduced, and seedlings are killed.

Identifying features: Yellow spots about 1.5 cm in diameter appear on the upper surface of leaves of apple and crab apple trees in the spring. Later, the spots enlarge, and dark-colored fruiting bodies are produced. Another type of fruiting body appears on the lower leaf surface in the summer; these fruiting bodies appear as whitish or tan columns and discharge yellow spores. Infection of the apple fruit occurs occasionally.

The small infected branches on eastern red cedar develop brown, spindle-shaped, warty or dimpled galls, up to 5 cm in diameter. During wet weather in the early spring of the second year, dimples on the galls rupture, and yellow-orange tendrils several centimeters long protrude; these tendrils are called telial horns.

Life history: The life cycle of the rust, which takes about 2 years, is completed on two different host species: eastern red cedar (18-20 months) and apple (4-6 months). In May, small, pale yellow spots appear on apple leaves as a result of infection by spores produced on eastern red cedar in the spring. These spots enlarge and become yellowish orange, with a red, band-like border. Tiny dark fruiting bodies that appear on these spots on the upper leaf surface in early summer are involved in fertilization. Another type of spore-producing body then develops as brownish-white columns on the leaf underside. The whitish cover of these fruiting bodies splits open, liberating yellow spores in late summer. These spores are dispersed by wind and infect needles and young stems of cedar. The fungus overwinters in eastern red cedar and produces galls the following spring or summer. In the second year of infection on cedar, after warm spring rains, spores are produced on yellow-orange, jelly-like telial horns. These spores germinate in place and produce another type of small spore, which infects the leaves of nearby apple trees.

Control: Foliage of apples can be protected by fungicide application during leaf development. Junipers can also be protected by midsummer application of fungicidal spray. Clipping or removal of young galls from cedar, or removal or avoidance of one of the hosts, is also recommended. Use of resistant varieties of apple or cedar is also suggested.

Additional information: Eastern red cedar, actually a juniper, is the only coniferous host known for the fungus in eastern Canada, but other junipers may also serve as hosts. Another fungus, *Gymnosporangium globosum* Farlow, produces symptoms similar to those of *G. juniperi-virginianae* on juniper, but the galls are smaller and the telial horns are shorter and more tongue-shaped. The alternate hosts for this rust are hawthorn and pear.

Selected bibliography

Boyce, J.S. 1961. Forest pathology. 3rd ed. McGraw-Hill Book Co., New York, NY. 572 p.

Parmelee, J.A. 1965. The genus *Gymnosporangium* in eastern Canada. Can. J. Bot. 43:239-267.

Prepared by Pritam Singh.

Plate 30

A. Fruiting bodies of *Gymnosporangium juniperi-virginianae*, the cedar-apple rust fungus, on cedar. Note the expanded spore horns.

B. Cedar-apple rust gall, caused by *Gymnosporangium juniperi-virginianae*, with expanded spore horns on the conifer host.

C. Cedar-apple rust gall, caused by *Gymnosporangium juniperi-virginianae*, on the juniper host as it is most often observed.

D. Globose gall rust gall, caused by *Gymnosporangium globosum*, with expanded spore horns on the conifer host.

E. Globose gall rust gall, caused by *Gymnosporangium globosum*, on the conifer host as it is usually seen.

F. Apple leaf infected by the cedar-apple rust fungus, *Gymnosporangium juniperi-virginianae.* The dark dots in the discolored spots are fruiting bodies of the spore stage involved in fertilization.

A

B

C

D

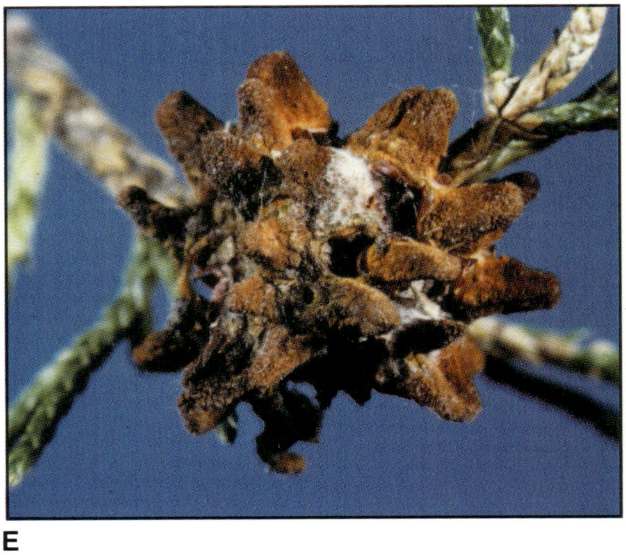

E

F

31. Pine needle rust
Coleosporium asterum (Dietel) H. Sydow & Sydow
Plate 31

Hosts: Mainly jack, red, and Scots pine; occasionally pitch pine; alternate hosts are various species of aster and goldenrod.

Distribution: Found throughout the range of its hosts in eastern Canada.

Effects on hosts: The current needles on pines are infected late in the season and are usually dead the following year. Some infected needles persist for 3 years. Because the current needles are not affected through most of the season, an additional stress must be involved for tree death to result. Defoliation can be a serious problem for Christmas tree and ornamental tree production, where loss of foliage reduces merchantability. There may be some growth reduction caused by the reduced amount of foliage.

Identifying features: In the spring, the infected area appears as a yellow spot or band with an orangish color in the center. The most readily recognized feature of this fungus is the white columnar blisters formed on the pine needles in late spring or early summer. Orange spores are usually evident beneath the white covering. Later in the season, this area on the needle is brown, with indications of tissue rupture. Infection of the alternate host is characterized by the development of orange, cushion-like masses on the undersides of affected leaves early in the summer. A similar reddish structure accompanies the orange masses later in the season.

Life history: *Coleosporium asterum* produces the five spore stages of a complete rust cycle. The first spore stage forms beneath the white columnar blisters that develop on the pine needles in late spring or early summer. The blisters rupture during wet weather, and the spores are released. The spores are wind-borne and cause infection of the leaves of the alternate hosts—aster and goldenrod. The fungus produces its second spore stage in the form of orange, cushion-like masses on the undersides of the infected leaves. This is often referred to as the repeating stage of the rust, as these spores infect other aster and goldenrod. These newly infected plants can also produce spores of the second stage and spread infection even further. Several generations of this stage may be produced during the course of summer. The third spore stage also forms on the undersides of the leaves of the alternate host, where it coexists with the second spore stage. These spores are formed in the late summer and appear as a reddish, cushion-like mass. They germinate in place, producing the fourth spore stage, which infects the pine needles where the fungus overwinters. Early in the spring, the fifth spore stage is produced as orange droplets on the now-apparent lesions of the fungus on the infected pine needles. These spores function in fertilization, and, later in the spring, the white blisters develop to repeat the cycle.

Control: Because the needles produced in the current year are not infected until late in the growing season, control is not necessary to prevent mortality. Control may be needed to prevent foliar damage to high-value plantation trees. Destruction of alternate host plants 300 m from planting sites by mowing or chemical means provides some control. The presence of asters and goldenrods should be avoided when selecting planting areas.

Additional information: Nicolls, van Arsdel, and Patton (see Selected bibliography) suggest that the *C. asterum* on red pine and jack pine alternates only with goldenrod, and the form on aster infects ponderosa pine. Consult their paper for an interesting account of their observations. *Coleosporium asterum* was previously called *Coleosporium solidaginis* Thüm. *Coleosporium viburni* Arthur is similar to *C. asterum* and alternates between species of *Viburnum* and jack pine. This species is not common.

Samples for diagnosis should include the rust pustules on the pine needles and a collection from the suspected alternate host, if possible.

Selected bibliography

Nicholls, T.H.; van Arsdel, E.P.; Patton, R.F. 1965. Red pine needle rust disease in the Lake States. U.S. Dep. Agric., For. Serv. Res. Note LS-58. 4 p.

Ouellette, G.B. 1966. *Coleosporium viburni* on jack pine and its relationship with *C. asterum*. Can. J. Bot. 44:1117-1120.

Prepared by H. Gross.

A

B

C

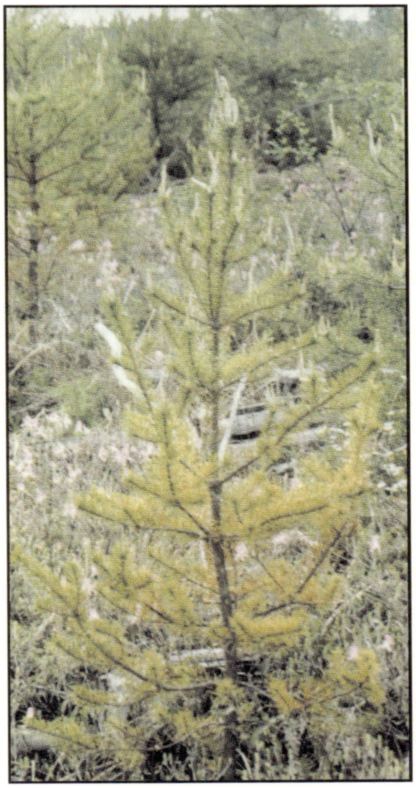

D

Plate 31

A. *Coleosporium asterum*, the causal organism of pine needle rust, fruiting on red pine needles.

B. Spore pustules of *Coleosporium asterum*, the causal agent of pine needle rust, on necrotic spots on pine needles.

C. Fruiting of *Coleosporium asterum*, the cause of pine needle rust, on goldenrod. The yellowish pustules are composed of the spores that infect other goldenrod, and the reddish pustules produce the spores that infect pine.

D. Jack pine infected by the pine needle rust fungus *Coleosporium viburni*. This needle rust fungus infects only jack pine needles and species of viburnum.

32. Spruce broom rust
Chrysomyxa arctostaphyli Dietel
Plate 32

Hosts: Mainly black, Norway, and white spruce; the alternate host is bearberry.

Distribution: Found throughout the range of its hosts in eastern Canada.

Effects on hosts: Infection of spruce stimulates bud formation, resulting in a tight proliferation of shoots and branches called a witches'-broom. Damage, which is primarily on the branches, can result in dead branches and spike tops. Trunk deformation and reduced growth can result from infection on the main stem. Death directly attributed to this disease is not common. Broken branches can open the tree to decay fungi and other organisms that may weaken the stem, leading to more significant damage as well as reducing merchantable volume. Many infected trees have significant cull because of deformation and breakage or death of tree crowns.

Identifying features: The most conspicuous feature of the disease is the broom with the etiolated needles. Late in the summer, the needles are also covered by orange spores, making the broom even more noticeable. These brooms are often visible from a considerable distance. The brooms shed their needles in the fall but are still rather obvious. In comparison, dwarf mistletoe brooms retain their needles throughout the year and also have a normal green color.

On bearberry, the fungus causes a purple leaf spot. In the spring, a spore stage on the undersurface of the bearberry leaf produces small, reddish-orange mounds or spots.

Life history: Spores produced on reddish spots on the undersides of bearberry leaves in the spring germinate in place and produce spores that infect the spruce. Spores are produced in pustules on the spruce in mid- to late summer. The covering of the pustules ruptures, releasing the orange spores, which carry the rust back to bearberry leaves, where the fungus overwinters. Unlike many other rust fungi, this fungus does not have a repeating stage to spread the disease to more bearberry plants. The rust fungus is systemic in the spruce brooms, so that the new needles on established brooms are infected every year. New brooms, however, must be initiated by the spores produced on bearberry. Broom and spore production usually does not occur on the spruce host until the year following a new infection. Brooms continue to live until the infected portion of the host is broken off or the host itself dies.

Control: The disease has not been sufficiently serious in eastern Canada to warrant control measures. Removal of diseased trees during thinning operations would contribute to control.

Additional information: The alternate host is also called kinnikinnick, and the rust also has the common name yellow witches'-broom. Red ring rot (*Phellinus pini* (Brot. : Fr.) A. Ames) (see Section 59) has been found to be the most common decay fungus colonizing tops and branches that were broken because of infection by *C. arctostaphyli.*

Selected bibliography

Savile, D.B.O. 1950. North American species of *Chrysomyxa.* Can. J. Res. (Sect. C) 28:318-330.

Ziller, W.G. 1974. The tree rusts of western Canada. Environ. Can., Can. For. Serv., Victoria, B.C. Publ. No. 1329. 272 p.

Prepared by D.T. Myren and H.L. Gross.

Plate 32

A. Witches'-broom and fruiting of *Chrysomyxa arctostaphyli*, the cause of spruce broom rust, on black spruce. Late summer fruiting gives the broom a yellowish color.

B. Witches'-broom of *Chrysomyxa arctostaphyli*, the causal agent of spruce broom rust, on black spruce. The proliferation of the shoots is evident.

C. Immature fruiting bodies of *Chrysomyxa arctostaphyli,* the cause of spruce broom rust, on black spruce. Some pustules have an orange cast, indicating they are almost ready to open and release the spores that will infect bearberry.

D. A leaf of bearberry, the alternate host for spruce broom rust, with a purple spot typical of infection by *Chrysomyxa arctostaphyli.*

A

B

C

D

33. Spruce needle rust

***Chrysomyxa ledi* (Alb. & Schwein.) de Bary and *Chrysomyxa ledicola* (Peck) Lagerh.**

Plate 33

Hosts: Mainly black, red, and white spruce; the alternate host is Labrador tea.

Distribution: Found throughout the range of its hosts in eastern Canada.

Effects on hosts: These two species of fungi infect the current year's needles and are responsible for most of the spruce needle rust in eastern Canada. The infected needles die, and the resulting defoliation, if severe, probably affects tree growth. Extensive stands of spruce with severe defoliation have been reported, but usually the amount of rust does not remain high for consecutive years. Normally, a considerable amount of healthy, older foliage remains and is able to sustain the trees. The rust fungi occasionally infect cones.

Identifying features: The white pustules and the orange to yellow spores that develop in them are evident on infected spruce foliage in midsummer and are the most conspicuous stage of the disease. The spores of *C. ledicola* are considerably larger than those of *C. ledi*, but the fungi are indistinguishable on spruce in the field.

Chrysomyxa ledicola is distinct on Labrador tea because it is the only rust fungus that fruits on the upper surface of the leaf. On Labrador tea, the rust fungi fruit on the foliage produced in the previous year.

Life history: Both rust fungi overwinter in the foliage of Labrador tea. In the spring, a spore stage develops that spreads the disease to other Labrador tea plants. Because this stage occurs every spring, the disease can persist on the alternate host in the absence of spruce. The reverse is not true, as both hosts are required for successful infection of spruce. In the early summer, a different spore stage develops on Labrador tea, and these spores spread the disease to the spruce host. Infection on spruce first appears as small reddish dots in which small fruiting bodies soon develop. These fruiting bodies produce spores involved in fertilization. Later, white pustules are formed under which the orange to yellow spores develop. Once these spores mature, the white cover tears open, and the spores are released; they are then carried by wind and rain splash to Labrador tea, where they initiate infection. These spores usually mature in the mid- to late summer. Both rust species can be found together on the same spruce host and even in adjacent pustules on the same needle.

Control: Chemical control of this disease on spruce does not seem necessary. Nurseries should not be located near swampy locations that typically contain considerable amounts of Labrador tea.

Additional information: *Chrysomyxa ledi*, which seems to be more common than *C. ledicola*, has a number of varieties. The most important in eastern Canada are var. *ledi*, which has Labrador tea as the alternate host, and var. *cassandrae* (Peck & G.W. Clinton) Savile, which has leather-leaf as the alternate host. Both *C. ledi* and *C. ledicola* can persist in Labrador tea far north of the range of spruce. *Chrysomyxa weirii* H. Jackson also occurs on spruce but does not have an alternate host. It produces spores in the spring on the 1-year-old needles.

Selected bibliography

Savile, D.B.O. 1950. North American species of *Chrysomyxa*. Can. J. Res. (Sect. C) 28:318-330.

Ziller, W.G. 1974. The tree rusts of western Canada. Environ. Can., Can. For. Serv., Victoria, B.C. Publ. No. 1329. 272 p.

Prepared by D.T. Myren and H.L. Gross.

A

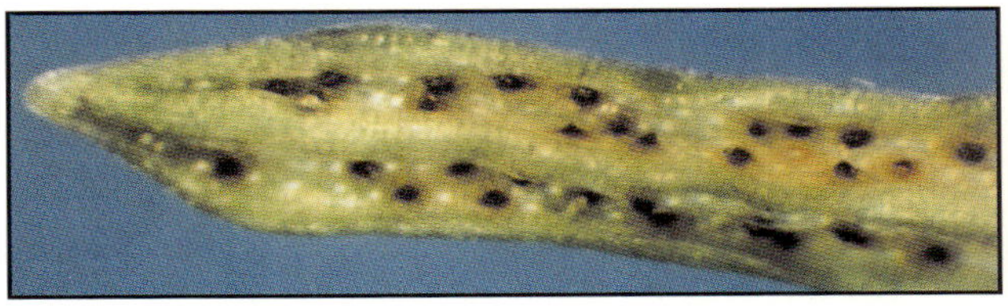

B

C

D

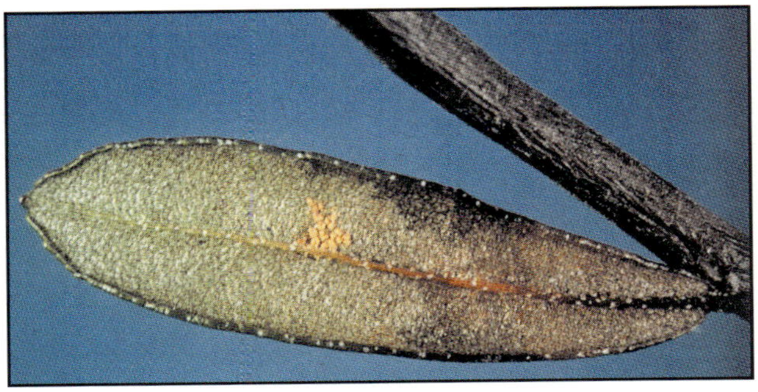

E

Plate 33

A. Spruce tree infection by *Chrysomyxa ledicola*, the cause of spruce needle rust.

B. Fruiting bodies of the fertilization spore stage on a spruce needle as is commonly observed for the spruce needle rust fungi *Chrysomyxa ledi* and *Chrysomyxa ledicola*.

C. Fruiting bodies of *Chrysomyxa ledi*, the cause of spruce needle rust, on a spruce needle. Note the orange color of the spore mass and the torn pustules, indicating the spores are ready for release.

D. Labrador tea with fruiting of the spruce needle rust fungus, *Chrysomyxa ledicola*.

E. Leather-leaf with fruiting of the spruce needle rust fungus, *Chrysomyxa ledi*.

34. Powdery mildew
Uncinula adunca (Wallr. : Fr.) Lév.
Plate 34

Hosts: Poplars and willows.

Distribution: Widely distributed throughout eastern Canada.

Effects on hosts: Individual leaves are killed by this fungus. Small trees can be damaged so severely that their growth rate is reduced. Mortality is not expected, although this disease could become a serious problem under conditions of high moisture and poor air circulation, possibly resulting in the death of some seedlings.

Identifying features: This fungus forms a white, superficial, cobweb-like growth on both surfaces of the infected leaves. During much of the summer, it looks like white velvet and can cover almost the entire leaf. Later in the summer, yellow or orange dots, which finally become black, are visible on the leaf surface. The black, pinhead-sized structures are the mature fruiting bodies, which have characteristic appendages on their surface; these can often be seen with a hand lens.

Life history: The black fruiting body of the fungus is the overwintering stage, and it is anchored to the fallen infected leaf by appendages. The function of the appendages is possibly to raise the fruiting body above the leaf surface to make it more subject to spread by wind and rain splash. In the spring, these fruiting bodies open and discharge their spores, which are then carried by air currents and cause the new infections on the young, healthy foliage. Once the leaf is colonized by the fungus,

A

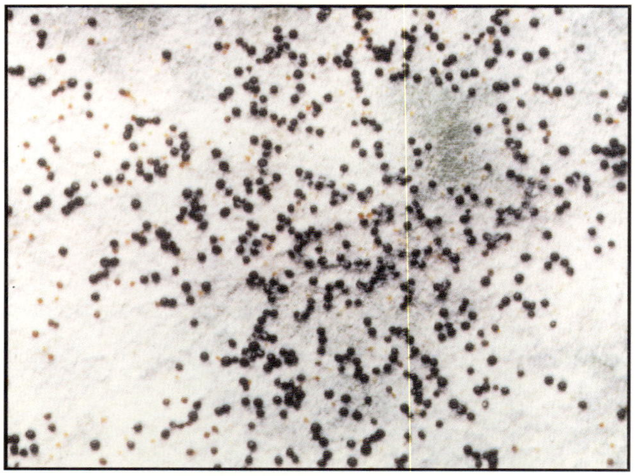

B

Plate 34

A. Powdery mildew, caused by *Uncinula adunca*, on a willow leaf. Some of the black fruiting bodies of the fungus are visible on the leaf.

B. Fruiting bodies of *Uncinula adunca*, the cause of powdery mildew, on a willow leaf. The black fruiting bodies are mature or near maturity, and the orange bodies are immature.

a second type of spore is produced. These spores are barrel-shaped and are formed in chains. They give the infected leaf a powdery or velvet-covered appearance. Wind and rain splash disseminate these spores, which initiate additional infections, thus intensifying the disease. Later in the summer, the overwintering spore stage begins to develop.

Control: Fungicides effectively control powdery mildew. Any cultural practice that improves air circulation in and around plants that suffer from this disease helps reduce infection. Evening watering of ornamentals should be avoided where powdery mildew has proven troublesome.

Additional information: This fungus was called *Uncinula salicis* (DC.) Winter and is found under this name in much of the early literature. The imperfect state is in the genus *Oidium*. There are six genera of powdery mildew, all of which are found in eastern Canada. They have a broad host range and are commonly seen on grass and herbaceous ornamentals as well as trees.

Selected bibliography

Boyce, J.S. 1961. Forest pathology. 3rd ed. McGraw-Hill Book Co., New York, NY. 572 p.

Parmelee, J.A. 1977. The fungi of Ontario. II. Erysiphaceae (mildews). Can. J. Bot. 55:1940-1983.

Prepared by D.T. Myren.

35. Apple scab
Venturia inaequalis (Cooke) Winter
Plate 35

Hosts: Mainly apple; occasionally hawthorn and mountain-ash.

Distribution: Common throughout eastern Canada.

Effects on hosts: This fungus attacks the leaves, blossoms, and fruits of its hosts. It causes discoloration, spotting, and distortion of the leaves and fruit, resulting in a reduction of yield and yield quality. Severe infection can also cause premature dropping of fruit, extensive defoliation, and mortality of young twigs. Three years of heavy defoliation on flowering crab results in significant twig dieback, and a single year of defoliation reduces the amount of foliage produced in the following year.

Identifying features: Infection of leaves by apple scab results in the development of circular brownish to gray spots, which change to olive green or almost black as the fungus develops. A dendritic pattern is often evident on the spots and is best seen with a hand lens. Early leaf fall and "shot holes"—holes in the leaf where dead material has fallen out—are also symptomatic of leaf infection. In some cases, infected leaves turn yellow, and the fungus appears on isolated green spots.

Infection on the fruit causes spots that vary in color from almost black to brown or gray, and distortion is common. Severe infection may result in cracking.

Life history: Fruiting structures of the fungus begin to develop on the fallen infected leaves in the late summer and fall, and they overwinter in an immature state. The cup-shaped fruiting bodies mature in the early spring following warm rains and discharge the spores that cause the primary infections. As these lesions develop, a second type of spore is produced, which can further spread and intensify the infection. Several generations of this spore state may develop during the season. Infection in all cases requires the extended presence of moisture, and dew alone will not suffice, although it may contribute. Rain is definitely needed, and the amount and time vary with the temperature. Apple scab is not as severe in very dry years as in wet years.

Control: Fungicides are very effective in the control of apple scab. Applications usually start shortly after bud break. Timing and number of sprays are very important and are related to the fungicide selected and weather conditions. In some areas, *V. inaequalis* has developed resistance to certain fungicides, but anyone interested in protecting ornamentals should have success by following manufacturers' recommendations for a fungicide marketed for scab control. Raking and destroying leaves in the fall to remove the overwintering state of the fungus are also recommended. Host varieties with resistance to scab should be considered when selecting trees for planting in areas where apple scab is a problem.

Additional information: The imperfect state of *V. inaequalis* is *Spilocaea pomi* Fr. : Fr. and is the state normally seen. The fruiting state that first occurs in the spring is not usually detected. Pressed leaves bearing the suspected lesions of the fungus make the best samples.

Selected bibliography

Anderson, H.W. 1956. Diseases of fruit crops. McGraw-Hill Book Co., New York, NY. 501 p.

Barr, M.E. 1967. The Venturiaceae in North America. Can. J. Bot. 46:799-864.

Prepared by D.T. Myren.

Plate 35

A. Apple scab, caused by *Venturia inaequalis,* on a leaf from flowering crab. Spores of the imperfect state of the fungus, *Spilocaea pomi*, will be abundant on the dark areas of fungus growth.

B. Fruit of flowering crab infected by the apple scab fungus, *Venturia inaequalis*.

A

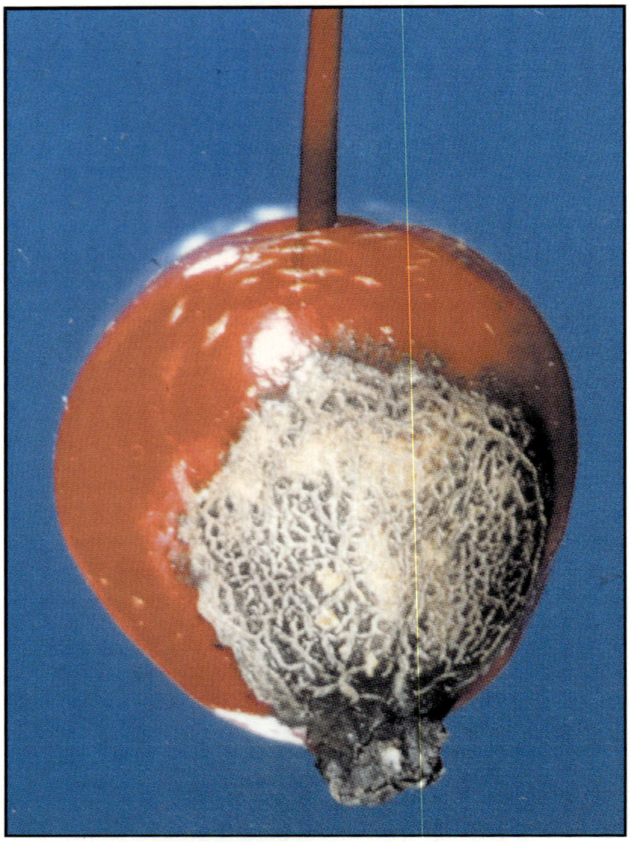

B

36. Willow scab and black canker of willow

***Venturia saliciperda* Nüesch and *Glomerella miyabeana* (Fukushi) v. Arx**
Plate 36

Host: Only willow.

Distribution: Occur in small, scattered patches throughout eastern Canada within the range of distribution of the host.

Effects on host: The fungus causing willow scab (*Venturia saliciperda*) infects young leaves, blackening and killing them rather quickly. The fungus causing black canker (*Glomerella miyabeana*) kills new shoots and causes cankers on the infected material. Defoliation may be complete and, if repeated for 2 or 3 successive years, is fatal to the tree; trees of any size may die in about 3 years. Cool, wet weather favors the disease, but the intensity of the disease varies from year to year and from locality to locality. Because willows are of little commercial value, the diseases are not considered important except where the host is used as a shade or an ornamental tree.

Identifying features: The most severe infections occur in the spring, often in the lower parts of the crown, and affect the young leaves; the mature leaves appear to be more resistant to infection. The fungus fruits on the lower leaf surface, forming olive-brown masses. Infected leaves are killed rapidly and become reddish brown or blackish, depending on the species of willow. The diseased leaves remain attached for some time, but they eventually dry up and fall off. The fungus moves from the leaves through the petioles and into the twigs, where it may cause some dieback and cankering. The fungus causing black canker comes later in the spring, killing leaves that may have escaped scab, as well as shoots and larger twigs. It also produces the typical black cankers. Small pink spots may be seen on the canker early in the summer, and these are soon followed by small black spots.

Life history: The fungus causing scab overwinters in a dormant state in the young twigs. It produces spores in the spring, which are dispersed by rain and cause infection of the new leaves.

The fungus causing black canker overwinters on the dead twigs and releases its spores 1–2 weeks after the scab fungus. Two spore types are produced: one appears as small pink spots and the other appears as small black spots. Both types of spores can be found in the early summer on the

Plate 36

Black canker, caused by *Glomerella miyabeana*, on young willow twigs.

infected twigs. They seem to spread the disease during the growing season. The small black spots are flask-shaped vesicles in which the spores are borne. This is the stage that overwinters.

Control: Control measures are largely applicable to ornamental and shade trees. They consist of pruning and destruction of diseased twigs and branches during the dormant season. These reduce the inoculum and prevent its further spread to healthy trees. Application of fungicidal sprays and use of resistant varieties of willows are also recommended.

Additional information: The imperfect state of *V. saliciperda* is *Pollaccia saliciperda* (Allescher & Tubeuf) v. Arx and is the state most commonly seen. It is almost always associated with *G. miyabeana*, which has *Colletotrichum crassipes* (Speg.) v. Arx as its imperfect state. *Glomerella miyabeana* is regarded as *Physalospora miyabeana* Fukushi by some mycologists who do not recognize the change made by von Arx. Because the two diseases caused by these fungi—scab and black canker—occur together, they are often considered as one disease. In the first one or two decades following the introduction of these fungi into North America from Europe, they killed thousands of willows in eastern Canada. At present, the disease seems to cause very little mortality, even in areas where the fungi are well established.

Selected bibliography

Boyce, J.S. 1961. Forest pathology. 3rd ed. McGraw-Hill Book Co., New York, NY. 572 p.

Davidson, A.G.; Fowler, M.E. 1967. Scab and black canker of willow. Pages 201-203 *in* A.G. Davidson and R.M. Prentice, eds. Important forest insects and diseases of mutual concern to Canada, the United States and Mexico. Dep. For. Rural Dev., Ottawa, Ont. 248 p.

Peace, T.R. 1962. Pathology of trees and shrubs with special reference to Britain. Oxford University Press, London. 723 p.

Prepared by Pritam Singh.

---

37. Snow blight
Phacidium abietis (Dearn.) J. Reid & Cain
Plate 37

Host: Balsam fir.

Distribution: Found throughout the range of balsam fir in Ontario and Quebec and on Campobello Island in New Brunswick.

Effects on host: Damage by *P. abietis* is most common in nurseries and on young, naturally regenerated stock under melting snow in the spring. It is most severe where snow is deep and slow in melting. The fungus spreads rapidly, and the damage often occurs in patches, particularly in nursery beds. Needles 1 year old and older are killed, but the buds are not harmed.

Identifying features: Completely brown foliage on branches below the snow line, often sharply delineated, is symptomatic. Affected seedlings in nurseries are often in well-defined patches, and an ephemeral cobweb-like mycelium may be seen covering them as the snow melts. The fruiting structures are formed in the fall on the undersides of needles killed by the fungus the previous spring. They are circular to oval, dark in color, and produced in rows, one on each side of the midrib. When mature, the undersurface of the needle over the fruiting bodies is ruptured, liberating the spores.

Life history: In late summer and fall, fruiting bodies on the infected needles mature and liberate spores during periods of fairly mild, moist weather. These wind-borne spores are responsible for primary infection. Spores landing on the surfaces of needles germinate and establish infection as soon as the needles are covered by snow. Needles of all ages are susceptible. Secondary infection occurs in the spring as mycelium from infected plants spreads to adjacent healthy foliage under melting snow. Once the snow is gone, spread of the fungus ceases. Small black microsclerotia may be produced on infected foliage. The microsclerotia may be involved in spread of the fungus, but their role is not completely understood.

Control: Fungicides applied in the early fall protect nursery stock and young regeneration. Susceptible tree species should not be grown in areas where snow tends to drift and melting is delayed. Infected seedlings and infected branches on windbreak trees should be removed and destroyed in the early summer.

Additional information: Much of the earlier information on snow blight is somewhat misleading because many of the fungal species involved had

A

B

C

D

E

Plate 37

A. Snow blight, caused by *Lophophacidium hyperboreum*, of blue spruce in a forestry nursery.

B. Mycelium of *Lophophacidium hyperboreum*, and spruce needles killed by this snow blight fungus.

C. Snow blight, caused by *Phacidium* sp., on spruce. Note the extensive damage to the needles below the snow line.

D. Snow blight, caused by *Phacidium* sp., on container stock at a forestry nursery. Snow cover remained over the lower portion of the stem for a prolonged period in the spring.

E. Fruiting of *Sarcotrochila balsameae*, the cause of snow blight, on the undersurfaces of balsam fir needles.

not been described. *Sarcotrochila piniperda* (Rehm) Korf and *Lophophacidium hyperboreum* Lagerb. are snow blights of spruce. *Phacidium infestans* P. Karsten, *Sarcotrochila balsameae* (J. Davis) Korf, and *Nothophacidium phyllophilum* (Peck) Smerlis cause snow blight of balsam fir. *Phacidium taxicola* Dearn. & House has been found on yew, and *P. infestans* also occurs on pine. Other species are known to cause snow blight, but the above-mentioned species are currently recognized as occurring in eastern Canada.

Samples of snow blight damage submitted for identification of the causal fungus should be collected in the fall and should contain mature fruiting bodies.

Selected bibliography

Hepting, G.H. 1971. Diseases of forest and shade trees of the United States. U.S. Dep. Agric., For. Serv. Agric. Handb. No. 386. 658 p.

Reid, J.; Cain, R.F. 1962. Studies of the organisms associated with "snow blight" of conifers in North America. II. Some species of the genera *Phacidium*, *Lophophacidium*, *Sarcotrichila* (sic) and *Hemiphacidium*. Mycologia 54:481-497.

Prepared by D.T. Myren.

---

38. Sooty mold
Catenuloxyphium semiovatum (Berk. & Broome) Hughes
Plate 38

Hosts: Basswood, elm, oak, and eastern white pine.

Distribution: Collected in central and southern Ontario.

Effects on hosts: Most of the sooty molds grow on honey-dew secreted by aphids and other sucking insects. The major effect of the sooty mold is physical coverage of the foliage, which reduces the area in which transpiration and photosynthesis can occur. This can reduce the vigor of a young tree, affecting growth and possibly survival. The scattered and relatively infrequent occurrence of sooty molds usually makes them a problem only on ornamental trees. Sooty mold has been serious in some balsam fir Christmas tree plantations in the Maritime provinces, resulting in degradation. The sooty mold followed an attack by the balsam woolly twig aphid.

Identifying features: Sooty molds are characterized by a black or brownish growth that covers the foliage and branches of the host. The black covering may be of varying thickness and shape.

Life history: Collections of *C. semiovatum* are limited, and its life cycle is not fully understood. Spores are found in fruiting structures during much of the summer. The fungus appears to spread by rain splash, but insects or wind may also be involved in long-range dissemination. Viable spores landing on surfaces where scale or aphid honey-dew is present initiate the establishment of the fungus. The fungus declines once the insect infestation ceases and may disappear within a few years.

Control: In most situations, control of the scales or aphids also controls the fungus. The unsightly nature of this fungus is of greatest concern on ornamentals and in Christmas tree plantations where insect control can be easily employed.

Additional information: Several fungi that cause sooty mold are found throughout eastern Canada and on a number of different hosts. The name *Fumago vagans* Pers. is often encountered in early literature, but this is now recognized as a mixture of *Cladosporium* and *Aureobasidium*. Some members of the genus *Capnodium*, a fairly well-known genus of sooty mold, have been collected in Ontario. The genus *Polychaeton* represents the imperfect state of a number of sooty molds and has also been collected widely in the province.

Selected bibliography

Hughes, S.J. 1976. Sooty molds. Mycologia 68:693-820.

Westcott, C. 1970. Plant disease handbook. 3rd ed. Van Nostrand Reinhold Co., New York, NY. 843 p.

Prepared by D.T. Myren.

Plate 38

A. Sooty mold on an oak leaf.

B. Sooty mold, caused by *Capnodium pini*, on eastern white pine needles.

C. Sooty mold, caused by *Scorias* sp., on an eastern white pine stem.

D. Sooty mold on jack pine branches and needles.

A

B

C

D

39. Godronia canker

Godronia cassandrae Peck f. sp. *betulicola* Groves

Plate 39

Hosts: Mainly white and yellow birch; occasionally speckled alder, trembling aspen, gray birch, and willow.

Distribution: Widespread in Quebec.

Effects on hosts: Mortality as high as 60% has been observed in natural regeneration of white and yellow birch. Cankers girdle branches and stems of saplings 0.5–5 cm in diameter and sometimes larger.

Identifying features: Swellings on branches or stems are usually associated with godronia canker. In summer, the disease can easily be detected by the wilting and browning of leaves on the infected saplings. In stands where the disease has been detected, apparently healthy branches or stems should be inspected for black areas on the usually brown bark; this discolored bark is the first symptom before wilting and browning. The fungus produces small fruiting bodies at the margin of the canker.

Life history: The life history of this fungus is not well known. However, tree inoculation tests have shown that the fungus produces one type of spore during the first growing season and a second type of spore the second year, for a 2-year life cycle. How or when the infection occurs on birch saplings is not known.

Control: No direct control is known for this disease.

Additional information: Godronia canker has been found frequently in natural forests and is also known to cause damage in birch plantations. Diaporthe canker, caused by *Diaporthe alleghaniensis* Arnold, is reported on yellow birch from the northeastern United States. It has symptoms similar to those of godronia canker but has been infrequently collected in eastern Canada.

Selected bibliography

Groves, J.W. 1965. The genus *Godronia*. Can. J. Bot. 43:1195-1276.

Smerlis, E. 1969. Pathogenicity of some species of *Godronia* occurring in Québec. Plant Dis. Rep. 53:807-810.

Prepared by G. Laflamme.

Plate 39

A. Dieback of yellow birch due to godronia canker caused by *Godronia cassandrae* f. sp. *betulicola.* Note cankers on the stem.

B. Fruiting of *Godronia cassandrae* f. sp. *betulicola*, the cause of godronia canker, on white birch.

A

B

40. Scoleconectria canker and dieback

Scoleconectria cucurbitula (Tode : Fr.) C. Booth

Plate 40

Hosts: Mainly eastern white pine; occasionally other conifers.

Distribution: Found throughout New Brunswick, Nova Scotia, Ontario, and Quebec.

Effects on hosts: This fungus causes cankers on branches and stems of trees in the natural forest and in plantations. The fungus kills small stems and branches by its girdling action and may also cause the dieback of branch tips.

Identifying features: Stem or branch cankers caused by *S. cucurbitula* are characterized by a relatively well-defined, depressed area of reddish bark. The infected bark remains attached to the tree and supports the development of fungal fruiting structures, which appear as small, red-brown spots less than 0.5 mm in diameter. The disease is usually first detected by the presence of dead foliage on a branch or a part of the tree. The cankers become evident later.

Life history: This fungus has long been considered a saprophyte, but recent pathogenicity tests have shown it to act as a parasite. Its life cycle is not well known, but two spore types have been identified. Our observations have shown that the invasion of the tissue by the fungus can be very fast at times but negligible at other times. This fungus has been found occasionally in association with white pine blister rust cankers.

Control: There are no known control measures for this particular disease, but, as with most cankers, it is suggested that infected branches and trees be cut to prevent new infections on the residual trees.

Additional information: This fungus is also known under the name *Zythiostroma pinastri* (P. Karsten) Höhnel. *Thyronectria balsamea* (Cooke & Peck) Seeler, which is similar to *S. cucurbitula*, is found mainly on balsam fir but also on pines, and it is considered to be a saprophyte. *Nectria macrospora* (Wollenw.) Ouellette and *Dermea balsamea* (Pack) Seaver occur on balsam fir and cause cankers that can lead to the death of the host. *Nectria macrospora* has been found only on the north shore of the St. Lawrence River, north of Baie Comeau, and on Anticosti Island. Removal and burning of infected branches and trees have considerably reduced infection centers of this fungus.

Selected bibliography

Booth, C. 1959. Studies of pyrenomycetes. IV. *Nectria* (Part I). Commonw. Mycol. Inst. (CMI) Mycol. Pap. 73. 115 p.

Smerlis, E. 1969. Pathogenicity tests of four pyrenomycetes in Québec. Plant Dis. Rep. 53:979-981.

Prepared by G.B. Ouellette.

A

B

C

D

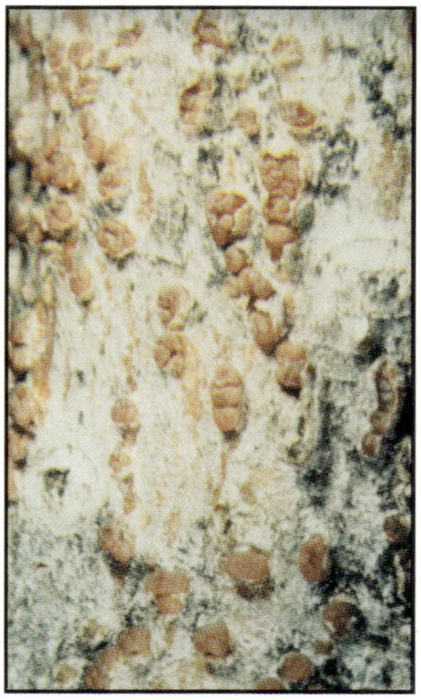

E

Plate 40

A. Scoleconectria canker, caused by *Scoleconectria cucurbitula*, on the stem of an eastern white pine.

B. Fruiting bodies of *Scoleconectria cucurbitula*, the cause of scoleconectria canker, on the stem of an eastern white pine.

C. Balsam fir damaged by *Thyronectria balsamea,* the cause of thyronectria canker.

D. Thyronectria canker, caused by *Thyronectria balsamea*, on the stem of a balsam fir.

E. Fruiting bodies of *Thyronectria balsamea*, the cause of thyronectria canker, on the stem of a balsam fir.

41. Nectria dieback
Nectria cinnabarina (Tode : Fr.) Fr.
Plate 41

Hosts: Mainly basswood and maples; rarely conifers.

Distribution: Widespread in eastern Canada.

Effects on hosts: This fungus is often found as a saprophyte on dead branches. It causes dieback of twigs and branches on trees under stress from wounds or other agents and is considered to be a weak parasite. On ornamental maples, it causes death of branches and, in severe infections, death of small trees. It has recently been shown to cause cankers on honey locust in the United States.

Identifying features: The most common and conspicuous signs of the disease are the cushion-shaped, light pink-orange fruiting bodies that appear on the bark. These fruiting bodies measure less than 1 mm in diameter and vary in number. The second type of fruiting body is a dark red color but is otherwise similar to the first type. The second fruiting body replaces the first type if the surrounding conditions are favorable for its development.

Life history: Infection occurs in the late winter or early spring through small wounds on the bark of branches and twigs. Once the fungus is established, it invades deeper into the bark and produces the pink-orange fruiting bodies. These structures produce spores, which cause new infections during the growing season. As the season progresses, the second type of fruiting body is produced. Spores produced in these fruiting bodies overwinter and are disseminated in the late winter or spring to start another cycle.

A

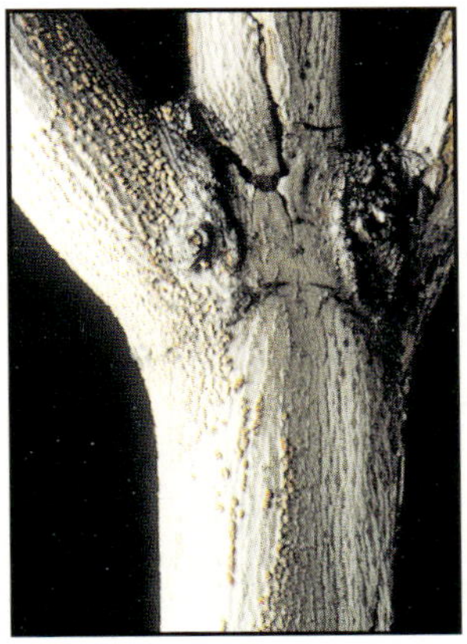

B

C

Plate 41

A. Nectria dieback of maple caused by *Nectria cinnabarina.*

B. Fruiting of *Nectria cinnabarina*, the cause of nectria dieback, on the stem of Norway maple. Note the bark cracking on the main stem killed by the fungus.

C. Fruiting bodies of *Nectria cinnabarina*, the causal agent of nectria dieback, on a Norway maple stem.

Control: Branches bearing fruiting bodies should be pruned and destroyed. Cankers on trunks may be removed if they are not extensive. To prevent infection, all dead branches should be removed, the tree should be protected against wounding, and tree species or provenances that are adapted to the climatic region should be planted.

Additional information: *Nectria cinnabarina* is often referred to as *Tubercularia vulgaris* Tode : Fr., the imperfect state of the fungus. The fungus has been identified on about 60 genera of woody plants, including a few conifers. Old fruiting structures of this fungus may be black and can be identified only by laboratory examination.

Selected bibliography

Hepting, G.H. 1971. Diseases of forest and shade trees of the United States. U.S. Dep. Agric., For. Serv. Agric. Handb. No. 386. 658 p.

Pirone, P.P. 1978. Diseases and pests of ornamental plants. 5th ed. John Wiley & Sons, New York, NY. 566 p.

Prepared by G. Laflamme.

---

42. Nectria canker
Nectria galligena Bresad.
Plate 42

Hosts: Largetooth and trembling aspen, basswood, beech, white and yellow birch, red and sugar maple, hybrid poplar, and willow.

Distribution: Widespread in hardwood stands of eastern Canada, although the amount of infection varies considerably between areas.

Effects on hosts: The major damage is the significant loss in volume as a result of large stem cankers affecting the butt log. Breakage occurs commonly at points of cankering, and this results in considerable reduction in merchantable volume. Cankers also serve as points of entry for decay fungi.

Identifying features: The older cankers are easily recognized by the absence of a bark covering and the presence of ridges of callus, the latter often being concentric or target-shaped. At times the bark, although loose, is retained and thus obscures the presence of a canker. Young cankers are not easy to detect. Most often they are associated with wounds or branch stubs and appear as flattened or depressed areas of the bark, which may be somewhat darker than normal. As the canker ages, some cracking of the bark or callus production is noted. Small, reddish, ovoid, inconspicuous fruiting bodies can often be found on the bark or wood on the margins of the canker.

Life history: Two types of fruiting bodies are produced by this fungus: one appears as tiny white pustules on recently killed bark of young cankers, and the other appears as reddish ovoid structures that develop in moist weather on the bark adjacent to the canker. The latter are much more common, and the spores they produce are probably more important in spread. The spores are spread by wind and rain splash in moist weather and initiate infection in wounds, branch stubs, leaf scars, and cracks at branch axils. Most new infections occur on trees under 20 years of age, which accounts for the large amount of damage to the butt log. Infection can also occur in branches. Infection does not occur through unwounded bark or dead tissue. Once established, the fungus colonizes the bark and moves a few millimeters into the wood tissue. The fungus grows slowly, and little damage occurs when the tree is actively growing. When growth of the trees slows or the tree becomes dormant, the fungus can spread more readily, liberating toxins that kill host tissue in advance of colonization. When tree growth resumes, the host produces a roll of callus in an attempt to seal off the wound. This cycle is repeated annually, with the consequent target pattern consisting of successive rolls of callus tissue.

Control: Control is accomplished primarily by removal of infected trees, shortened rotations, or regeneration to less susceptible species. Young cankered trees should be removed during thinning or improvement cuttings. When older, severely cankered trees are present, they should also be felled. If infection levels are below 20% of the species selected for regeneration in a mature stand, normal management can be followed. If infection levels are over 20%, regeneration to more resistant tree species should be considered. The latter practice is also suggested for young stands with 50% or more of the crop trees infected. Sanitation cuttings during stand improvement and shortened rotations are suggested in young stands when infection levels are under 50%. Successful eradication of the fungus from a woodlot is not possible because it can exist as a saprophyte and also because some young infections can be missed or may be at a symptomless stage.

Additional information: Nectria canker is also referred to as European canker, but the fungus is suspected as being native to both continents. A similar fungus, *Nectria cinnabarina* (Tode : Fr.) Fr. (see Section 41), is seen frequently on ornamental maples and produces small, spherical, red fruiting bodies that are larger than those produced by *N. galligena. Nectria cinnabarina* causes death of branches and young trees, but typical target cankers are not formed.

Samples should include fruiting structures when possible.

Selected bibliography

Boyce, J.S. 1961. Forest pathology. 3rd ed. McGraw-Hill Book Co., New York, NY. 572 p.

Hepting, G.H. 1971. Diseases of forest and shade trees of the United States. U.S. Dep. Agric., For. Serv. Agric. Handb. No. 386. 658 p.

Prepared by D.T. Myren.

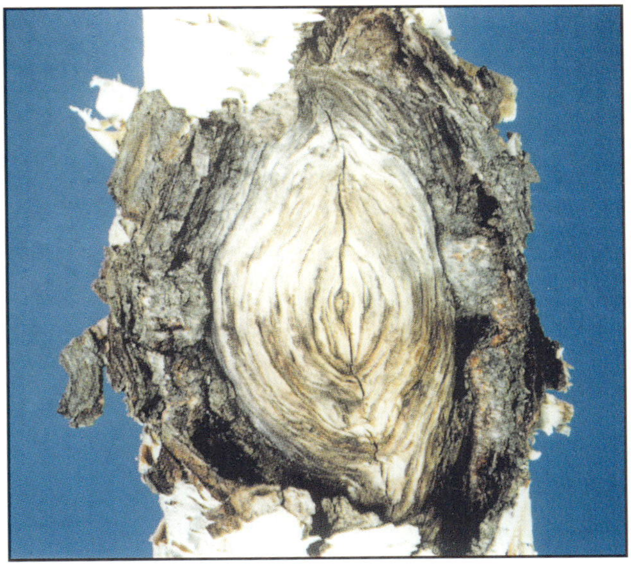

A

B

Plate 42

A. Nectria canker, caused by *Nectria galligena*, on a white birch exhibiting the typical target pattern.

B. Fruiting bodies of *Nectria galligena*, causal agent of nectria canker, on the margin of a canker on white birch.

43. Shoot blight
Gremmeniella laricina (Ettl.) Petrini *et al.*
Plate 43

Hosts: Tamarack and European and western larch.

Distribution: Only recorded from Quebec; found most often in Laurentides Reserve and north of Lake Albanel.

Effects on hosts: Mortality of young trees 1–3 m in height is associated with the development of cankers on the main stem. The infection is usually limited to the 1- and 2-year-old twigs of young trees. The rate of mortality is usually higher in a plantation than in a natural forest.

Identifying features: During the growing season, the disease is detected by brown needles on the new shoots. Late in the summer, dead twigs are evident. A depression in the bark or a shrinkage of

A

B

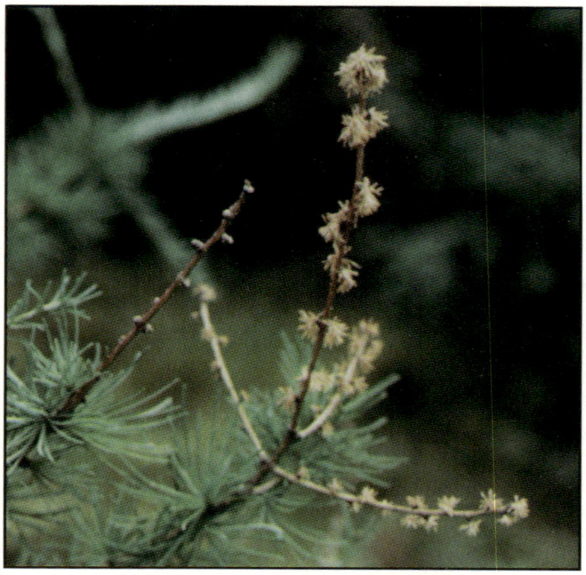

C

Plate 43

A. Canker caused by *Gremmeniella laricina*, the shoot blight fungus, on tamarack. Note fruiting bodies on the canker.

B. Fruiting body of the shoot blight fungus, *Gremmeniella laricina*, on a tamarack branchlet.

C. Dead shoots on tamarack symptomatic of shoot blight caused by *Gremmeniella laricina*.

a twig develops between the healthy and the diseased tissue. Later on, minute (<1 mm in diameter), dark brown fruiting bodies are usually present.

Life history: The spores usually mature in summer and are dispersed before the end of the growing season. New infections are then initiated in the current year. The infected twig produces fruiting bodies the following year. Another type of spore may also be produced in the fall, but it is rare.

Control: No control measure is known for this disease, but fungicides applied during the period of spore discharge should provide adequate protection. Control is advised only for high-value plantations and nurseries.

Additional information: This fungus was earlier described as *Scleroderris laricina* (Ettl.) Gremmen. After a recent revision, it was classified in the new genus *Encoeliopsis* and then moved into the genus *Ascocalyx* by Schläpfer-Bernhard. Recently, it was moved into the genus *Gremmeniella*, where we place it in this work. This disease, previously reported from Switzerland and recently described in British Columbia, is apparently quite new in eastern Canada, but it seems to be well established. The imperfect state of the fungus is *Brunchorstia laricina* Ettl. This fungus is closely related to *Gremmeniella abietina* (Lagerb.) Morelet (see Section 47), which causes scleroderris canker.

Selected bibliography

Funk, A. 1981. Parasitic microfungi of western trees. Environ. Can., Can. For. Serv., Pac. For. Res. Cent., Victoria, B.C. Inf. Rep. B.C.-X-222. 190 p.

Müller, E.; Dorworth, C.E. 1983. On the discomycetous genera *Ascocalyx* Naumov and *Gremmeniella* Morelet. Sydowia Ann. Mycol. 36:193-203.

Petrini, O.; Petrini, L.E.; Laflamme, G.; Ouellette, G.B. 1989. Taxonomic position of *Gremmeniella abietina* and related species: a reappraisal. Can. J. Bot. 67:2805-2814.

Prepared by G. Laflamme.

44. European larch canker
Lachnellula willkommii (R. Hartig) Dennis
Plate 44

Hosts: Mainly tamarack, although all species of the larch family are susceptible.

Distribution: Found only in southern New Brunswick and mainland Nova Scotia.

Effects on hosts: The fungus causes cankers on branches and stems of infected trees. Multiple cankering is common, with as many as 20 on a single branch. Infected stems are deformed and contain excess resin. When girdled, the portion of the tree or branch distal to the canker dies.

Identifying features: Young cankers appear as swellings on twigs and branches or as depressions on larger stems and are accompanied by exuding resin. This gives the cankers a shiny appearance, often with a bluish hue. White, hairy, cup-shaped fruiting bodies with yellowish interiors are usually found in or around the canker during most of the year. Needles above the canker on affected branches and small stems either shrivel up and die in the spring or discolor early in the fall. The cankers are perennial and enlarge from year to year.

Life history: Infection is most likely by spores through short shoots in the late summer and fall when the tree is becoming dormant. The fungus progresses into the branch or stem while the tree's resistance mechanisms are lowered. The following year, the formation of the canker begins, accompanied by foliage symptoms either in the spring or in late summer. Fruiting body production commences early in the cankering stage and continues as the perennial canker increases in size.

Control: Control is practical only for high-value trees, such as nursery, plantation, and ornamental trees. Pruning or shearing affected parts reduces spore production. Thinning provides better air movement through plantations, lowers humidity, and thus may lower spore release. Fungicides are known to prevent infection.

Additional information: The fungus was first found in Massachusetts in 1927, having been introduced from Europe in infected nursery stock. Early control efforts seemed successful in eradicating the fungus, but it reappeared in 1935 and again in 1952. Each reappearance of the disease was followed by rigorous sanitation. It was first found in Canada in 1980. The approximate age of cankers can be determined in cross section from the number of deformed annual rings.

In early literature the fungus was called *Dasyscypha willkommii* (R. Hartig) Rehm, and in some European literature it is referred to as *Trichoscyphella willkommii* (R. Hartig) Nannf.

Samples for diagnosis should include the canker and the area between green and discolored portions of the branch.

Selected bibliography

Boyce, J.S. 1961. Forest pathology. 3rd ed. McGraw-Hill Book Co., New York, NY. 572 p.

Buczaki, S.T. 1973. Observations on the infection biology of larch canker. Eur. J. For. Pathol. 3:228-232.

Prepared by L.P. Magasi.

A

B

Plate 44

A. Damage to tamarack shoots by European larch canker caused by *Lachnellula willkommii.*

B. Fruiting bodies of *Lachnellula willkommii* on a canker produced by the fungus on tamarack.

45. Eutypella canker of maple
Eutypella parasitica R.W. Davidson & R.C. Lorenz
Plate 45

Hosts: Mainly sugar maple; commonly red maple; occasionally black, Manitoba, Norway, and silver maples.

Distribution: Generally distributed throughout the range of maple in Ontario and Quebec; not yet reported from the Maritime provinces or Newfoundland.

Effects on hosts: The disease causes mortality by girdling trees less than 12 cm dbh (diameter at breast height, or 1.3 m above the ground). On larger trees, the canker is perennial and becomes an entry point for decay, thus making the tree susceptible to wind breakage. As 90% of the cankers occur at less than 3.7 m from the ground, the decay and trunk malformation render the first log practically useless for plywood or saw timber. Girdling seldom occurs on ornamental trees, as trees usually grow fast, and infection occurs later than in natural stands. The cankers, however, make the trees more susceptible to wind breakage, thus making them hazardous to the community and reducing their life span.

Identifying features: This disease causes a canker, usually on the trunk of the tree, and is characterized by a relatively well defined area of dead bark surrounded by a bulge of callus tissues. On red maple, the bulge may be less pronounced than on other maples, appearing as irregular swellings around the dead bark area. In 80% of the cases, a dead branch stub or its scar is found near the center of the canker. The dead bark always remains attached to the tree, and, after a few years, long-necked fruiting bodies develop in the bark in scattered or clustered black spots. Neat cuts in the bark tissue, made with a sharp blade, reveal the black, rounded fruiting chamber at the base of the necks where the spores are produced.

Eventually, decay fungi infect the wood under the canker. One of these (*Oxyporus populinus* (Schumacher : Fr.) Donk)) commonly produces a typical white fleshy fructification, partly covered with green moss and located more or less in the center of older cankers.

Another means of identifying *E. parasitica* requires the removal of pieces of bark at the upper or lower ends of the canker at the margin of healthy and infected bark and examination of this area for the presence of a pale beige or cream mycelial fan that is produced by the fungus.

Life history: Fruiting bodies develop in dead bark between 3 and 5 years after infection. Spores are released from the black fruiting structures whenever the temperature is above 4°C and the bark has been wetted by rain. Infection is believed to occur in branch stubs. In a maple stand, spread of the spores rarely exceeds 100 m from the infected points.

Control: Removal of sporulating cankers should reduce the likelihood of infection of healthy trees, particularly in uneven-aged or regenerating stands where old cankered trees occur among young maples. Trees 12 cm dbh and over and bearing a canker should be thinned out, as their crop value will probably be nil, and removal will leave growing space for healthy trees. On ornamental trees, pruning the lower branches is recommended to prevent the formation of rough, natural branch stubs that can serve as infection sites.

Additional information: It is estimated that a harvested maple infected by *E. parasitica* would lose about 12% of its total wood volume. As the area of the tree infected would be the butt log, the total merchantable loss would be nearly 50%. This includes the wood lost to decay fungi commonly associated with the canker.

Selected bibliography

Kliéjunas, J.T.; Kuntz, J.E. 1974. Eutypella canker, characteristics and control. For. Chron. 50:106-108.

Lachance, D. 1971. Inoculation and development of eutypella canker of maple. Can. J. For. Res. 1:228-234.

Prepared by D. Lachance.

A

B

C

D

Plate 45

A. Eutypella canker on sugar maple caused by *Eutypella parasitica.*

B. Young eutypella canker on maple, showing stem deformation and black fruiting area of the causal fungus, *Eutypella parasitica.*

C. Fruiting structures of *Eutypella parasitica*, the causal agent of eutypella canker, on sugar maple. The upper portion of the fruiting structures has been removed to expose the tubes through which spores move to reach the surface.

D. Decay fungus *Oxyporus populinus* fruiting on a eutypella canker.

46. Fire blight
Erwinia amylovora (Burrill) Winslow *et al.*
Plate 46

Hosts: Mainly apple, and American, European, and showy mountain-ash, and pear; occasionally cotoneaster, crab apple, hawthorn, plum, and spiraea.

Distribution: Generally distributed throughout eastern Canada.

Effects on hosts: Death may occur in a few weeks following infection, especially in young, highly vigorous trees. Usually trees die in one or two growing seasons. Sometimes only infected branches die, and the tree recovers.

Identifying features: Affected blossoms and leaves wilt and collapse. Leaves rapidly turn brown but remain on the tree. The appearance and spread of these leaf symptoms on mountain-ash are quite rapid and spectacular. The bark of branches and stems becomes reddish and water-soaked at the advancing edge of the infection but later turns black. Whitish droplets may appear on dead bark, but they dry and darken a few days later. Cracks usually appear soon on dead bark, and the wood underneath infected bark turns black.

Life history: The disease is caused by a bacterium. It overwinters in the living bark, at the outer edge of diseased stems and branches. During warm and humid spring weather, droplets of exudate containing the bacterium ooze from the bark. Splashing rain, birds, and insects then spread the bacterium to healthy new twigs and branches on the same tree or to other trees. Infection may occur through stomata, lenticels, or wounds. New infections can occur at any time during the growing season, but the periods of greatest susceptibility are at blossom time and when rapid and succulent twig growth occurs.

Control: A close watch, especially from early to midsummer, to detect early development of symptoms is the key to control. When symptoms are found, the infected twigs must be cut out at least 30 cm below the last sign of infected bark. The pruning tool should be disinfected after each cut, by wiping it clean with a cloth soaked in either methyl alcohol (wood alcohol) or household bleach diluted into nine parts of water. Treatment in dry weather is advised, as bacterial production and release are lower then. The removed infected parts should be destroyed to prevent them from serving as a source of infection.

Examination of susceptible trees during fall or winter facilitates the detection of cankers (dead and discolored bark). Also, contamination and spread during removal of infected parts are very much reduced at this time.

At blossom time, bactericidal sprays (antibiotics) may help to prevent infection. Sprays should be repeated immediately after bloom. Preventive fungicides are also reported to be effective in control.

Succulent growth favors development of the disease; thus, overfertilization (especially with nitrogen), late fertilization, overwatering, and severe pruning to increase branching should be avoided when possible.

Additional information C-E.A. Winslow was chairman of a committee on bacterial nomenclature established by the Society of American Bacteriologists. The *et al.* following his name refers to the members of that committee.

Selected bibliography

Pirone, P.P. 1972. Tree maintenance. 4th ed. Oxford University Press, London. 574 p.

Tattar, T.A. 1978. Diseases of shade trees. Academic Press, New York, NY. 361 p.

Prepared by D. Lachance.

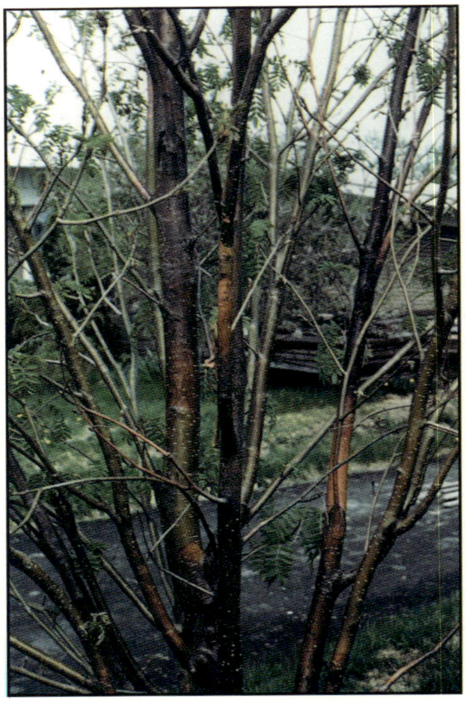

A

B

Plate 46

A. Mountain-ash stems with necrotic bark and cankers typical of fire blight caused by the bacterium *Erwinia amylovora.*

B. Early stage of fire blight, caused by the bacterium *Erwinia amylovora*, on mountain-ash shoots.

47. Scleroderris canker
Gremmeniella abietina (Lagerb.) Morelet
Plate 47

Hosts: Mainly Austrian, jack, red, and Scots pine; rarely eastern white pine and black and white spruce.

Distribution: North American race found in New Brunswick, Ontario, Quebec, and a few areas in Nova Scotia; European race found in New Brunswick, Newfoundland, Ontario, and Quebec.

Effects on hosts: The North American race of *G. abietina* infects young trees and rarely kills trees over 2 m tall. Branches in the lower crown of larger trees may become infected and are important because the fungus may persist there and spread to younger trees. The European race of the fungus causes damage to trees of all ages and sizes and, under favorable climatic conditions, may spread rapidly to the upper crown.

Infection takes place at branch tips, and the fungus grows back toward the main stem, usually killing one internode per year. The terminal of the main stem can also be infected. Once the fungus reaches the stem, it grows around it, and that portion of the tree at and above the point of girdling quickly dies. If the fungus does not successfully girdle the tree but does kill a portion of the bark, a canker results. Extensive mortality has been noted in both young plantations and naturally regenerated stands. Older, cankered trees that successfully outgrow the fungus may have a volume and quality reduction in the butt log because of weakening and pitch accumulation in the cankered area.

The European race of the fungus progresses more rapidly and infects entire branches at any height in a tree, regardless of its age. When environmental conditions are favorable for the fungus, large trees can be killed in a few years' time, often without any canker being formed.

Identifying features: The most obvious symptom of infection is the browning of the basal portion of the second-year needles, which occurs in late spring. This symptom may last up to 1 month on long needles but progresses until the entire needle is brown. The infected shoots die the same year the needle symptoms are evident. If the North American race is present, small brown fruiting bodies often develop abundantly in the spring on the material killed the previous year. These structures open in wet weather, forming a small, cup-like structure exposing a whitish-gray surface. They close and shrink as the weather dries. A second fruiting structure, also brown and somewhat spherical, forms on infected material of the current or previous year but is often difficult to find. This second fruiting structure is usually the only one observed when the European race is present. Stem cankers with a green stain under the bark are also characteristic of the disease. Unthrifty trees with numerous shoots killed back in areas where the disease is known are also symptomatic.

Life history: With the American race, fruiting structures produced on material killed the previous year open, forming cup-like structures in moist weather in spring and early summer. When open, a whitish surface in which spores are formed is exposed, and the spores are discharged. These spread by wind and rain splash, causing infection in new shoots. The following year, this shoot infection becomes evident on what are then the second-year needles. A second spore stage is produced in the bark of current infected tissue in early spring, and spores ooze from these in wet periods. In midsummer, these spores are produced in small brown spheres formed on the bark of current or previously infected tissue. These also open in moist weather, and the spores are spread primarily by rain splash. This causes an intensification of the infection.

The second spore stage, which is spread mainly by rain splash, predominates on trees infected with the European race. As this race is more virulent, infection occurs not only on the new shoots but on the entire branch.

Control: Nursery infection can be significantly reduced by removing infected trees in the windbreaks, particularly infected branches in the lower crown of larger trees. Sanitation in plantations may be beneficial if replanting is being considered. The disease is most severe in low areas within a plantation, and less susceptible species might be selected for planting in these locations.

Clipping and removal of infected branches have been successful in controlling only the beginning of an infection in plantations. Pruning all trees in infected plantations can be effective if the rate of infection is low. Other control measures are currently under study. A preventive measure in regions where scleroderris canker is present involves a survey to detect the disease in new plantations and the removal of infected material every 2 or 3 years. As trees become older, pruning the lower branches (about one-third of the crown) reduces infection, especially on highly susceptible species such as red pine. Fungicides can be used effectively in nurseries and Christmas tree plantations.

The European race is under quarantine law; this means that no transportation of trees or branches out of plantations infected with the European race is permitted. This is a fairly effective means of slowing the spread of this race of the fungus, as airborne spores are not an important source of infection.

Additional information: This fungus was known as *Scleroderris lagerbergii* Gremmen and *Ascocalyx abietina* (Lagerb.) Schläpfer-Bernhard and is found under these names in the literature. The imperfect state of the fungus is *Brunchorstia pinea* (P. Karsten) Höhnel.

Samples for diagnosis of this disease should have fresh needle symptoms or evidence of fruiting. The canker alone may be suitable. It is possible to make a diagnosis of recently infected material by searching in the laboratory for "cryptopycnidia," a fungal structure produced in the bark of shoots. This is useful in nurseries for the inspection of seedlings collected before their spring lifting date.

Two races of the fungus, the North American and the European, have been identified in Canada. The European race is usually more virulent, but it can be difficult to differentiate these races in the field, especially in young plantations. A serological test performed in the laboratory is currently the

A

B

C

D

E

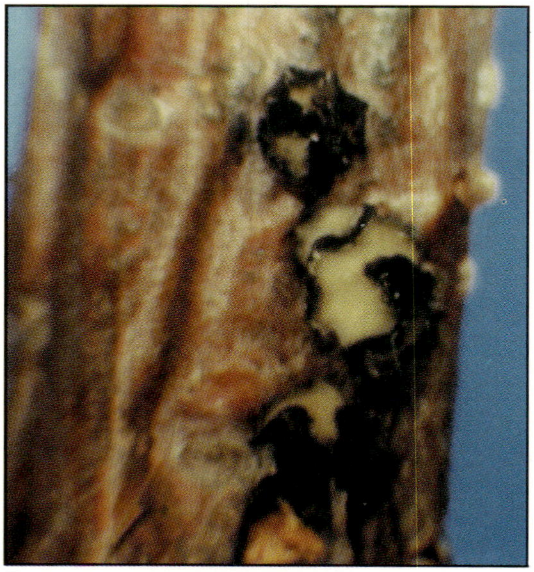
F

only reliable method available to identify races of this fungus.

Cenangium ferruginosum Fr. : Fr. is frequent on dead pine branches and, when dry, resembles the shrunken fruiting structure of *G. abietina.* The disk in open fruiting structures of *C. ferruginosum* is yellowish in color, whereas in *G. abietina* it is white.

Selected bibliography

Dorworth, C.E.; Davis, C.N. 1982. Current and predicted future impact of the North American race of *Gremmeniella abietina* on jack pine in Ontario. Environ. Can., Can. For. Serv., Great Lakes For. Res. Cent., Sault Ste. Marie, Ont. Inf. Rep. 0-X-342.18 p.

Laflamme, G. 1991. Scleroderris canker on pine. For. Can. Que. Reg. Inf. Leaflet LFC 3. 12 p.

Manion, P.D., ed. 1984. Scleroderris canker of conifers. Forestry sciences. Martinus Nijhoff/Dr. W. Junk, Boston, MA. 273 p.

Prepared by D.T. Myren and G. Laflamme.

Plate 47

A. Red pine infected by the European race of *Gremmeniella abietina*, the causal agent of scleroderris canker.

B. Jack pine with canker typical of the North American race of scleroderris canker caused by *Gremmeniella abietina*.

C. Early foliar symptoms of infection on red pine by *Gremmeniella abietina*, the cause of scleroderris canker. Note the brown needle bases.

D. Fruiting bodies of the perfect state of *Gremmeniella abietina*, the causal agent of scleroderris canker, on a jack pine stem.

E. Fruiting bodies of *Brunchorstia pinea*, the imperfect state of *Gremmeniella abietina*, the cause of scleroderris canker.

F. Fruiting bodies of the perfect state of *Cenangium ferruginosum*, cause of cenangium canker, on a Scots pine stem. Note the yellow interior of the fruiting structure, compared with the white color of *Gremmeniella abietina*.

48. Ceratocystis canker
Ceratocystis fimbriata Ell. & Halsted
Plate 48

Host: Trembling aspen.

Distribution: Widespread in Quebec.

Effects on host: The canker can kill small trees; on larger aspen, the fungus spreads slowly and is usually active for many years without killing the host. Many cankers on the same tree eventually girdle it, bringing about its death. Decay fungi can infect the host through open wounds created by perennial cankers, thus rendering the tree susceptible to wind breakage.

Identifying features: Ceratocystis canker is usually diagnosed by the target-shaped canker made up of concentric ridges. The ridges are composed of a callus tissue formed by the reaction of the tree to the fungus. Each year the fungus breaches the callus barrier the tree had created the previous year, so the tree responds by creating a new barrier, which again is breached. This can go on for many years and results in the target-shaped canker. The dead bark surrounding the canker is black, as are the fruiting bodies of the fungus.

Life history: There is little information on the life cycle of this fungus. Infection is believed to occur at branch stubs, which are common at canker centers. Infection may occur on leaves and petioles, and the fungus then invades twigs through leaf scars. The fungus could also enter through wounds. *Ceratocystis fimbriata* produces spores in the spring, and insects may be involved in the spread and transmission of the fungus.

Control: Like most hardwood cankers, new infections can be reduced by removing infected aspens, but this is not usually commercially feasible. Ceratocystis canker is known to infect only aspen in Canada, so the disease will not spread to other, more valuable trees in the stand.

Additional information: At least nine other species of *Ceratocystis* have been reported on aspen, but *C. fimbriata* is more often associated with the target-like canker. Nectria canker caused by *Nectria galligena* Bresad. (see Section 42) is very similar to ceratocystis canker, and examination of fruiting bodies of these fungi is necessary to make an accurate identification. Black fruiting bodies are difficult to find on a black background, and samples should be brought to a laboratory for a more critical examination.

Selected bibliography

Hinds, T.E. 1972. Ceratocystis canker of aspen. Phytopathology 62:213-220.

Manion, P.D. 1981. Tree disease concepts. Prentice-Hall, Englewood Cliffs, NJ. 399 p.

Manion, P.D.; French, D.W. 1967. *Nectria galligena* and *Ceratocystis fimbriata* cankers of aspen in Minnesota. For. Sci. 13:23-28.

Prepared by G. Laflamme.

A

B

Plate 48

A. Trembling aspen infected by the canker fungus *Ceratocystis fimbriata*.

B. Ceratocystis canker on trembling aspen caused by *Ceratocystis fimbriata*. Note the branch stub in the center of the canker.

49. Dothichiza canker of poplar
Cryptodiaporthe populea (Sacc.) Butin
Plate 49

Hosts: Mainly hybrid, Lombardy, and white poplar.

Distribution: Widely distributed in eastern Canada.

Effects on hosts: This disease is found on young, newly planted trees and in nurseries. Established young trees under stress and large ornamental trees can also be severely attacked — many branches die, and the trees become very unsightly. Stem cankers usually result in weakened hosts or mortality, so that breakage by winds or snow loads is likely. Branch cankers may result in breakage and may also serve as entry points for infection by decay fungi.

Identifying features: The cankers first appear as slightly sunken areas of bark, often developing around the base of twigs and branches. Bark color in the infected area may be darker than normal. Cracks in the bark may also be noted once the fungus has girdled the stem and the host dies. The fruiting bodies develop in the spring as dark, pinhead-sized structures on the dead bark.

Life history: The fruiting structures of the fungus are formed on dead bark throughout the growing season. The spores are issued in creamy- to amber-colored tendrils through a small pore at the top of the fruiting body, or they may ooze out, collecting around the apex. They are spread by rain splash and probably by birds, insects, and movement of infected stock. Infection occurs only through wounds, leaf scars, and bud scale scars and is most successful when bark moisture and turgor are lowest. This situation arises during the winter months and after the growth of the tree has ceased or slowed down in the fall. It may also occur if planting stock is allowed to dry excessively. The fungus will grow at colder temperatures than will the host, and canker development often occurs when the tree is dormant. Infections during the spring and summer are often sealed off by callus tissue if the host is growing vigorously. Fall infections, possibly through leaf scars caused by premature leaf drop, are suspected as being the most successful in establishing the fungus in the host and initiating cankers. The fungus overwinters as spores in unopened, late-developing fruiting bodies and also within the infected tissue.

Plate 49

Dieback of poplar hybrid due to infection by *Cryptodiaporthe populea*, the causal agent of dothichiza canker.

Control: Control of this disease is usually attempted through cultural measures. Dense stocking and the excessive use of nitrogen fertilizer should be avoided. Foliage disease, which causes premature leaf fall, should be controlled when possible. Planting must be done carefully and on proper sites. Pruning should not be done late in the growing season, and all pruning wounds must be clean. Older plantings should be thinned to reduce stand density. Desiccation of planting stock should be prevented. All these measures are part of proper tending and aid disease control. In addition, trees and branches that are cankered should be removed when possible. Fungicides can be used in a control program, but careful growing practices should be able to bring the disease to an acceptable level.

Additional information: The imperfect state of the fungus causing dothichiza canker was known as *Dothichiza populea* Sacc. & Briard for many years and can be found under this name in early literature. The imperfect state of the fungus is now called *Discosporium populeum* (Sacc.) B. Sutton and is the state found in eastern Canada. The perfect state, *C. populea*, has not been recorded in North America. There is evidence suggesting that this disease was European in origin.

Samples should include fruiting bodies where possible, but well-developed cankers may be adequate.

Selected bibliography

Hepting, G.H. 1971. Diseases of forest and shade trees of the United States. U.S. Dep. Agric., For. Serv. Agric. Handb. No. 386. 658 p.

Waterman, A.M. 1957. Canker dieback of poplars caused by *Dothichiza populea*. For. Sci. 3:175-183.

Prepared by D.T. Myren.

50. Septoria leaf spot and canker *Mycosphaerella populorum* G.E. Thompson Plate 50

Hosts: Eastern cottonwood, introduced poplar, and introduced poplar hybrids; occasionally native poplar and native poplar hybrids.

Distribution: Widespread in most of eastern Canada but not yet found in Newfoundland.

Effects on hosts: Damage is usually light in natural stands. The disease is more serious in nurseries and plantations where leaf infection may be severe and may cause premature defoliation. Cankers formed by the fungus provide an entrance for other canker-producing fungi and decay fungi that weaken the stem. Canker and leaf spots have been reported on introduced poplars, introduced hybrid poplars, and cottonwood. Native poplars seem to be affected only by the leaf spot form of the disease.

Identifying features: Leaf spots have an irregular shape and are usually brown with a darker margin. The color is quite variable in intensity and darker on the upper leaf surface. The diameter of the spot varies from 1 to 15 mm, and a larger diseased area is often formed by coalescence of many smaller spots. Fruiting structures develop on the lower surface of the lesion and appear as small black dots. Septoria leaf spot resembles other leaf spot diseases, and it is usually necessary to examine the spores under the microscope for proper identification. A branch or stem canker is recognized by a depressed area and a change in color of the bark, especially at the margin where it is orange-brown to black.

Plate 50

A. Septoria canker caused by *Mycosphaerella populorum* on hybrid poplar.

B. Septoria leaf spot on balsam poplar caused by *Mycosphaerella populorum.*

C. Balsam poplar defoliated by *Mycosphaerella populicola*, a causal agent of septoria leaf spot.

D. Septoria leaf spot on white birch caused by *Septoria betulae.*

A

B

C

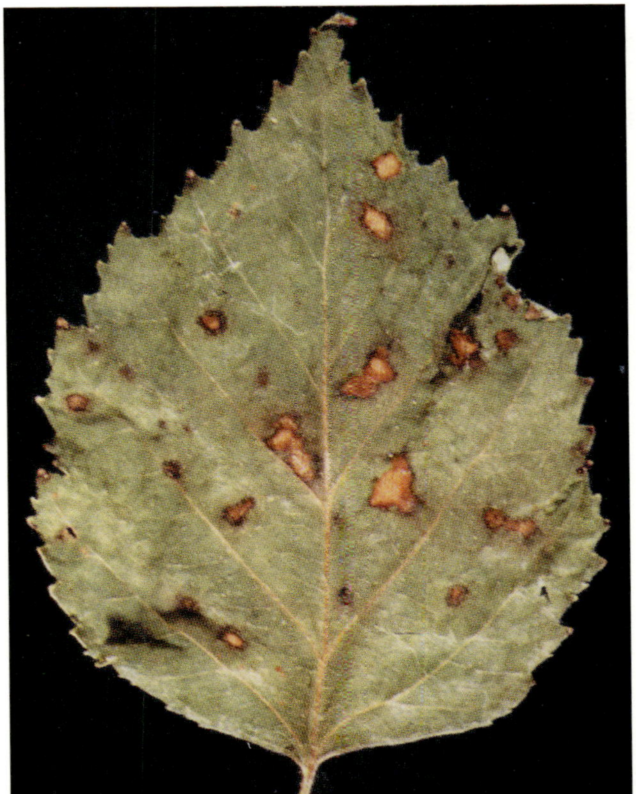

D

Life history: The fungus infects the leaves in the spring and produces spores throughout the summer. These spores develop in structures embedded in the leaf and are issued in whitish to pinkish tendrils. They can cause further stem and leaf infections. The fungus continues to develop on fallen infected leaves and overwinters on this dead material. In the spring, a second type of spore is produced, released during wet weather, and carried by the wind or rain splash to leaves, branches, or stems, where new infections can occur and the life cycle is continued.

Control: Control measures are recommended only for nurseries and plantations. Prevention is the best measure and is accomplished through sanitation, which reduces infection in nurseries. Sanitation must be rigorous and include the plowing under or removal of all material that could harbor the fungus and clearcutting of all native poplars in or near the nursery. In plantations, only clones resistant or less susceptible to the disease are recommended. Because the fungus can infect cuttings and persist during storage, all cuttings should be inspected for infection before planting to avoid the introduction of the disease into new areas.

Additional information: *Mycosphaerella populorum* is the perfect state of *Septoria musiva* Peck. A similar fungus that is found on balsam poplar, *Septoria populicola* Peck, or its perfect state *Mycosphaerella populicola* G.E. Thompson, is also common in eastern Canada. This fungus has caused heavy defoliation of balsam poplar in several large areas of Ontario, almost on an annual basis. On birch, a septoria leaf spot caused by *Septoria betulae* Pass. is found frequently. Particularly severe damage was reported on birch in northern Ontario in 1984.

Selected bibliography

Boyce, J.S. 1961. Forest pathology. 3rd ed. McGraw-Hill Book Co., New York, NY. 572 p.

Thompson, G.E. 1941. Leaf-spot diseases caused by *Septoria musiva* and *Septoria populicola*. Phytopathology 31:241-254.

Prepared by G. Laflamme.

51. Hypoxylon canker
Hypoxylon mammatum (Wahlenb.) P. Karsten
Plate 51

Hosts: Mainly trembling aspen; occasionally speckled alder, largetooth aspen, red and sugar maple, and balsam poplar; rarely white and yellow birch and willow.

Distribution: Widely distributed throughout most of eastern Canada but has not been reported from Newfoundland.

Effects on hosts: Hypoxylon canker is a serious disease of trembling aspen, often causing extensive mortality. The fungus becomes established in the inner bark and grows both vertically and horizontally in the stem, killing the bark in the areas it colonizes. Eventually the stem is completely girdled, and that portion distal to the canker dies. In many cases, the infected trees break at the canker even before girdling by the fungus is completed.

Identifying features: The first symptom observed is a yellowish-orange discoloration of the bark, often surrounding a branch stub or dead branch. The following year the canker enlarges, the orange-yellow discoloration continues to demarcate the boundary, and the bark shows blistering and cracking. Gray, pillar-like structures are found under the blistered bark. In the third year, small patches of hard, gray, raised fruiting structures are observed in the area where pillar-like structures were observed the previous year. A white mycelial fan may be found under the bark. The cankers often appear almost black from a distance, may be up to 1 m in length, and can encircle the entire stem.

Life history: The first spore stage is produced approximately 2 years after infection, with the spores being borne on pillars that push up and

A

B

C

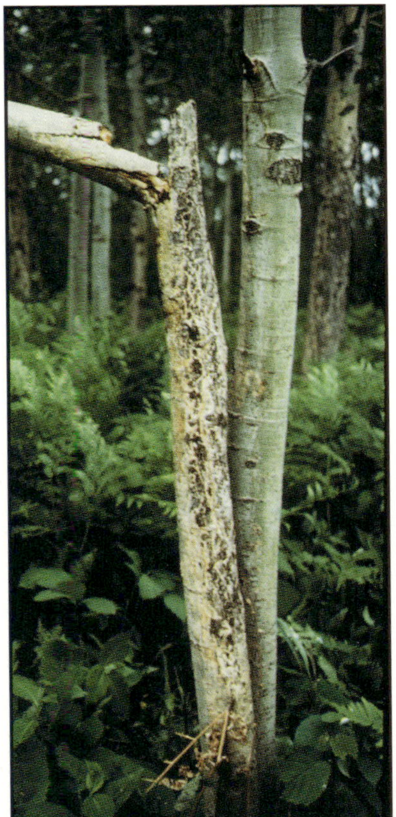

D

Plate 51

A. Trembling aspen infected by *Hypoxylon mammatum*, the cause of hypoxylon canker. Note the yellowish color of the bark on the canker margin.

B. Pillars produced by *Hypoxylon mammatum* on an infected trembling aspen. The pillars cause the raising and splitting of the bark typical of hypoxylon canker.

C. Fruiting structures of the perfect state of *Hypoxylon mammatum*, the cause of hypoxylon canker.

D. Breakage of trees infected by *Hypoxylon mammatum* is a common feature associated with hypoxylon cankers.

rupture the dead bark. These spores are not known to cause infection but may have a role in fertilization. The second type of fruiting body is produced in the third year in the area where the pillars were observed. The spores produced in these structures are ejected and carried by the wind and initiate the new infections. Wounds and branch stubs have long been suspected as points in which infection can occur. Recent evidence suggests that insect wounds and galls, particularly those caused by the poplar gall borer (*Mecas inornata* Say), serve as important infection sites.

Control: Silvicultural techniques form the basis for reducing losses due to hypoxylon canker. A dense stand and closed canopy are unfavorable to the fungus. Stands where infection exceeds 25% should be harvested promptly. Lower levels of incidence allow the period before harvest is required to be prolonged. Some aspen clones are more susceptible and should not be allowed to regenerate. Because infected trees often break at the point of cankering, it is recommended that severely cankered trees be removed in areas where breakage could result in damage to people or property.

Additional information: This fungus was known in early literature as *Hypoxylon pruinatum* (Klotzsch) Cooke and the species on maple as *Hypoxylon blakei* Berk. & M.A. Curtis.

Samples for diagnosis of this fungus should include the second spore stage.

Selected bibliography

Anderson, R.L. 1964. *Hypoxylon* canker impact on aspen. Phytopathology 54:253-257.

Berbee, J.G.; Rodgers, J.D. 1964. Life cycle and host range of *Hypoxylon pruinatum*and its pathogenesis on poplars. Phytopathology 54:257-261.

Boyce, J.S. 1961. Forest pathology. 3rd ed. McGraw-Hill Book Co., New York, NY. 572 p.

Prepared by D.T. Myren.

52. Cytospora canker
Valsa sordida Nitschke
Plate 52

Hosts: Mainly largetooth and trembling aspen, balsam, black, hybrid, and Lombardy poplar, and willow; occasionally eastern cottonwood, Norway, silver, and sugar maple, and mountain-ash; rarely pin cherry.

Distribution: Widely distributed in eastern Canada.

Effects on hosts: *Valsa sordida* is a weak parasite that causes damage only when the hosts are under stress. The heaviest infections occur in young trees, but severe damage is also caused to cuttings in nurseries, in plantations, and in winter storage. Losses of up to 75% of ornamental poplars have been reported in some nurseries.

Valsa sordida is known to cause branch dieback and even death of small stems. Small branches may be killed quickly. True cankers rarely develop because infected trees are so weakened by other agents that healing tissues are not formed. The diseased area enlarges in size until the stem is girdled. Severely attacked trees die within 2 to 3 years.

Plate 52

A. Poplar infected by *Valsa sordida*, the cause of cytospora canker. Note the orange fruiting of the fungus near the center of the developing canker.

B. Spore tendrils on willow, characteristic of fruiting of the imperfect states of both *Valsa* and *Leucostoma*.

C. Cytospora canker on trembling aspen caused by *Leucostoma nivea*. Note the branch stub in the center of the canker and the fruiting of the fungus on the dead bark.

D. Fruiting bodies of *Cytospora nivea*, the imperfect state of *Leucostoma nivea*. The dark dot in the center of the fruiting bodies is the pore from which the spores will exude.

A

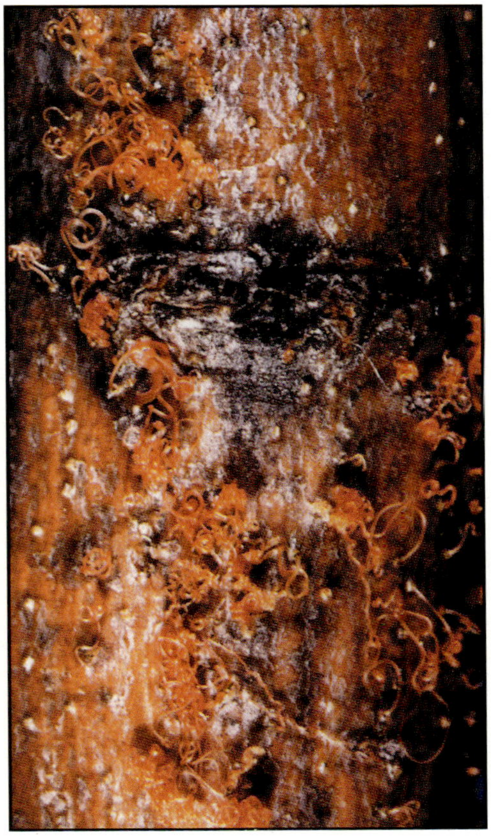

B

C

D

Identifying features: The disease first appears as a necrotic area of bark, often around a small wound. It may also appear as a brown, circular to oval, sunken lesion on the limbs or trunk. In either case, initial infection is followed by death of patches of bark under which discolored, watery, foul-smelling wood is often found.

Life history: The fungus enters the tree through wounds on dead or weakened branches and causes the death of an area of bark. Small erumpent fruiting bodies of this fungus appear on the dead bark as pinhead-sized pimples with a flat, grayish-black top. During moist periods, spores are exuded in long, bright yellow to orange tendrils or threads from a pore in the center of the top of the fruiting structure. The spores are spread by rain splash or carried by wind, birds, or insects to other trees, where new infections may occur.

Another reproductive stage of this fungus develops on dead stems but is relatively rare in North America. In the fall, the fruiting structures of this stage develop deep in the bark and produce long necks, which push through the bark surface. The spores move up the neck and are spread by wind, birds, or insects. This stage is usually found only on trembling aspen and, because it is rare, presumably does not play an important role in the dissemination of the fungus.

Control: Because this fungus attacks weakened trees, the most important control measure is maintaining good host vigor through applications of fertilizers and water when needed and avoidance of wounds. Pruning of diseased branches and removal of cankers also control the spread of the fungus. Particular care should be given to wound prevention and the prevention of water loss during storage of cuttings. Some hybrid poplars are more resistant to the disease and should be selected for plantations.

Additional information: The imperfect state of *V. sordida* is *Cytospora chrysosperma* (Pers. : Fr.) Fr. The closely related fungus *Leucostoma nivea* (Hoffm. : Fr.) Höhnel is also common on poplars. The imperfect state of *L. nivea* is *Cytospora nivea* (Hoffm. : Fr.) Sacc. *Cytospora nivea* can be distinguished from *C. chrysosperma* by its deep reddish spore tendrils and the very white disk from which they are issued. Samples for diagnosis should include the fruiting structures if possible.

Selected bibliography

Boyce, J.S. 1961. Forest pathology. 3rd ed. McGraw-Hill Book Co., New York, NY. 572 p.

Christensen, C.M. 1940. Studies on the biology of *Valsa sordida* and *Cytospora chrysosperma*. Phytopathology 30:459-475.

Hepting, G.H. 1971. Diseases of forest and shade trees of the United States. U.S. Dep. Agric., For. Serv. Agric. Handb. No. 386. 658 p.

Prepared by D.T. Myren.

53. Cytospora canker
Leucostoma kunzei (Fr. : Fr.) Munk
Plate 53

Hosts: Mainly balsam fir, larch, and blue spruce; occasionally eastern white pine and black, Norway, red, and white spruce.

Distribution: Found in New Brunswick, Nova Scotia, Ontario, and Quebec.

Effects on hosts: Infection usually occurs on the lower branches of the host and results in premature branch death. The disease is most often found on trees more than 20 years old, although younger trees can be attacked. The disease spreads throughout the tree, destroying its value as an ornamental. Trunk cankers are not common, but they do occur and can kill the host.

Identifying features: The first symptom of the disease is an off-green color of the foliage of individual branches. An examination of the branch back towards the stem reveals an area of heavy resin exudation. Some branch swelling may occur at that site, although the heavy resin accumulation tends to obscure this. The needles eventually turn brown and are shed. The fruiting bodies can be found by carefully peeling back the bark around the resinous area. Usually one major pore in a grayish disk is evident, at the base of which are a number of roughly circular chambers radiating out from the center. A black line surrounding the fruiting body is often evident. On old cankers, the inactive fruiting structures can be recognized by a black margin and a dingy gray powdery center. The radiating chambers are just barely or not at all discernible.

Life history: The fungus overwinters in fruiting bodies embedded in the bark of the infected host. Spores are exuded in yellow-orange tendrils and can be produced at any time during the normal growing season when wet conditions are present. The spores are spread by rain splash and to some extent by birds and insects, which may account for long-distance spread. Infection occurs through wounds. The fungus appears to be more successful on trees under stress, although healthy vigorous trees can be infected. A second type of fruiting structure is also produced but is not often encountered.

Control: The best method of control of cytospora canker is removal of the infected branches as soon as any evidence of the disease is noted. Pruning tools used should be wiped carefully with rubbing alcohol between cuts when several branches are to be removed. Pruning should not be done during wet weather, as removal of branches when spores are present might assist in the spread. Pruning other infected trees in the area can be helpful by reducing possible sources of spore production.

Additional information: The imperfect state of *L. kunzei* is *Cytospora kunzei* Sacc. and is the state of the fungus usually found. In early literature, *L. kunzei* is referred to as *Valsa kunzei* Fr. Three varieties of *L. kunzei* are recognized, with different hosts for each.

Samples submitted for identification should include a portion of a branch with resin accumulation, plus several inches on each side.

Selected bibliography

Jorgensen, E.; Cafley, J.D. 1961. Branch and stem cankers of white and Norway spruce in Ontario. For. Chron. 37:394-400.

Waterman, M.A. 1955. The relation of *Valsa kunzei* to cankers on conifers. Phytopathology 45:686-692.

Prepared by D.T. Myren.

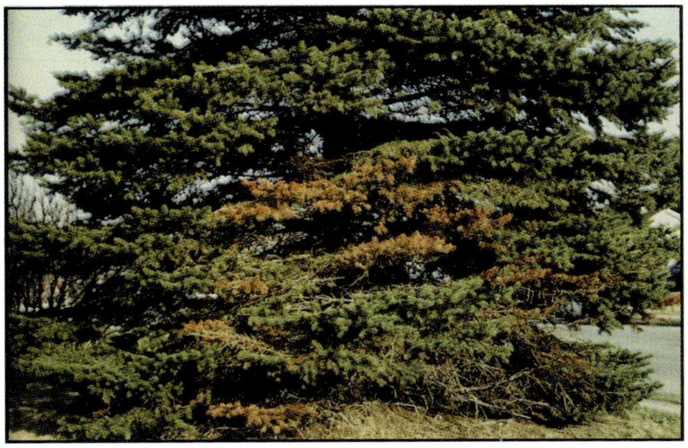

A

B

Plate 53

A. A spruce tree with scattered dead branches typical of infection by *Leucostoma kunzei*, the cause of cytospora canker on this host.

B. Cytospora canker on spruce with heavy resinosis typical of infection by the causal fungus, *Leucostoma kunzei*.

54. Beech bark disease
Nectria coccinea (Pers. : Fr.) Fr. var. *faginata* Lohman, Watson & Ayers
Plate 54

Hosts: American and European beech.

Distribution: Widely distributed in Nova Scotia, Prince Edward Island, most of New Brunswick, and southern Quebec.

Effects on hosts: The disease is a "complex" involving scale insects followed by the fungus, and it causes extensive cankering and deformation of the stem, rendering the trees unsuitable for any use other than as firewood. Most trees eventually succumb to the disease either directly or as a result of attack by secondary organisms.

Identifying features: Varying sizes of white fluffy "wool" covering tiny yellowish insects on the stem are indicative of active insect attack, during or in advance of canker formation. Severely infested trees, with or without cankers, may appear covered with snow. Cankers are pockmarks of varying sizes on the stem, encircled by ridges formed as the tree attempts to wall off infection. Cankers often coalesce on severely affected trees. Fruiting bodies of the fungus appear as clusters of small, deep red, flask-shaped structures in and around the cankers.

Life history: The scale insect overwinters on the bark in a partly immature stage and covered by the white, wool-like wax. It matures about midsummer when it is seen as a spherical, yellow, legless "glob." It produces eggs asexually (there are no males). Crawlers emerge from the eggs, move about or are dispersed by wind, then settle down, pierce the living bark with their feeding tubes, and secrete the woolly wax. The insects, once settled, do not move. They can survive only on living bark.

The spores of the fungus are spread by wind and rain splash. They gain entry to the tree through bark injuries made by the scale insect. Once established, the fungus kills the bark and forms white, cushion-shaped structures that give way to the red, flask-shaped or somewhat lemon-shaped fruiting bodies. Fruiting bodies always appear in clusters and are present 3–5 years after the first invasion by the insect. Although they are evident from early spring to late fall, they are most conspicuous during or just after moist periods.

Control: Practical control is possible only for ornamental trees and only against the insect. The tree can be sprayed to kill the scale insect, or the "wool" can be removed either by scrubbing with a solution of detergent or by using a strong stream of water from a garden hose. In either case, care should be taken not to injure the bark of the tree. In forest situations, the spread of the disease can be slowed by vigorous selective cutting and removal of infested and infected stems in the early stages of an outbreak.

Additional information: The scale insect, *Cryptococcus fagisuga* Lindeman, is known to occur in Ontario, although beech bark disease has not yet been found in that province. Another fungus on beech stems, *Ascodichaena rugosa* Butin, causes rough, black, circular patches on the trunk or branches but no injury to the tree. The dark fruiting bodies break through the bark surface, are produced in groups, and open by a slit. *Nectria galligena* Bresad. (see Section 42) is also a common colonizer of wounds caused by the beech scale.

Selected bibliography

Boyce, J.S. 1961. Forest pathology. 3rd ed. McGraw-Hill Book Co., New York, NY. 572 p.

Ehrlich, J. 1934. The beech bark disease: a *Nectria* disease of *Fagus*, following *Cryptococcus fagi* (Baer.). Can. J. Res. 10:593-692.

Prepared by L.P. Magasi.

Plate 54

A. Beech affected by the beech bark disease caused by *Nectria coccinea* var. *faginata.*

B. Fruiting bodies of *Nectria coccinea* var. *faginata*, the causal agent of beech bark disease.

C. The black fruiting bodies of *Ascodichaena rugosa*, which is a common fungus on beech bark but is not considered a serious problem.

A

B

C

55. Caliciopsis canker
Caliciopsis pinea Peck
Plate 55

Hosts: Balsam fir and eastern white pine.

Distribution: Reported only on balsam fir in Quebec and occasionally on eastern white pine in Ontario.

Effects on hosts: The reaction of the host to the fungus is a characteristic roughening of the bark. On eastern white pine, it may also cause a canker clearly delimited by the striking host response. The invasion of the fungus is said to be superficial and limited to the bark, so the impact of the disease is minimal even if the canker is perennial.

Identifying features: The rough bark of a portion of the trunk is usually conspicuous. Hair-like fruiting structures are typically present and readily visible in the center of the individual lesions.

Life history: The fungus is generally considered to be a parasite of wounds. It is believed that the fungus penetrates directly through old lenticels on eastern white pine and balsam fir, although it has been demonstrated that caliciopsis spores enter the bark of other tree species in areas damaged by insects. Once the spores have germinated and penetrated the host tissues, the fungus can establish itself and continue to produce spores year after year. It is suspected that the fruiting bodies mature in late winter and spring, but they are visible all year round, with different stages of development present on a single specimen.

Control: Trees are not necessarily threatened by the disease, but it is recommended that infected trees be removed when a stand is being thinned.

Additional information: *Caliciopsis pinea* is often called cork bark disease. On eastern hemlock, a new species, *C. orientalis* Funk, has been described; another species, *Caliciopsis calicioides* (Ell. & Ev.) Fitzp., has been reported on living bark of poplars.

Selected bibliography

Boyce, J.S. 1961. Forest pathology. 3rd ed. McGraw-Hill Book Co., New York, NY. 572 p.

Funk, A. 1963. Studies in the genus *Caliciopsis*. Can. J. Bot. 41:503-543.

Prepared by G. Laflamme.

Plate 55

Cork bark disease, caused by *Caliciopsis pinea*, on balsam fir.

56. Eastern dwarf mistletoe
Arceuthobium pusillum Peck
Plate 56

Hosts: Mainly black spruce; occasionally red and white spruce and tamarack.

Distribution: Common throughout the range of its hosts in eastern Canada.

Effects on hosts: Eastern dwarf mistletoe is one of the most damaging parasites of black spruce. The damage includes reduction in growth (height and diameter) and consequent loss in volume, reduction in wood quality, and reduction in cone and seed production. It is the major cause of reduced stocking in black spruce stands; in heavily affected areas, the stocking level is so low that commercial harvesting is not economically feasible.

Eastern dwarf mistletoe attacks trees of all ages and sizes, killing young saplings as well as heavily infected mature or overmature trees.

Identifying features: The most apparent symptoms of eastern dwarf mistletoe infection are stimulated growth and phototropic response of branches at the point of infection, resulting in the production of swollen, bushy, distorted, compact masses of branches and twigs, known as "witches'-brooms." These brooms usually persist and often ooze sap for as long as the host remains alive and may grow to be 1–3 m in diameter. Heavily infected trees with older infections have severely malformed branches and spiked tops. Hand-lens examination of a cross section of the infected area of a branch shows the wedge-shaped sinkers.

Life history: Eastern dwarf mistletoe has separate male and female plants and reproduces through seeds. The life cycle of *A. pusillum* requires at least 4 years from infection of the host to the first crop of fruit. Male and female plants are located on separate branches or on separate trees. Flowers appear in the spring of the third year and are insect-pollinated. Following fertilization, female plants produce green to dark brown, one-seeded berries. Each mature fruit contains a viscous-coated seed that is forcibly ejected on maturity, usually in late summer. The seeds readily adhere to needles or other plant parts on which they happen to land and stay there until rain moistens the viscous coating, after which the seeds slide down and become attached to twigs. This is followed by germination on the bark the next spring. The radicle is formed by germination and moves along the branch until it encounters a bud or leaf base. At that point, it produces a holdfast, which forms a haustorium on its lower surface. This primary haustorium penetrates the bark to the cambium, and a cortical haustorium then develops in the living cambial tissue. Sinkers originate from the cortical haustoria and become embedded in the host as new growth of the host develops. Aerial shoots begin to form after 2 years, mature the following year, and flower in the spring of the fourth year, with mature fruit formed by that fall.

Control: Because eastern dwarf mistletoe is an endemic parasite, it persists unless eradicated. Control can be accomplished through sanitation, which consists of cutting or pruning infected trees, usually at harvesting time. Treatment depends on the host, age of the stand, severity of infection, locality, and economics. The incidence of the parasite and the losses caused by it can be significantly reduced by clearcutting the infected areas plus a surrounding 20-m-wide isolation strip and checking areas treated for eastern dwarf mistletoe 10 years after treatment, to make sure the disease is under control.

Additional information: Weakened trees are more susceptible to windthrow, drought, and attack by insects and fungi. Tissue killed by eastern dwarf mistletoe provides entrance points for stain and decay fungi.

Witches'-brooms produced by eastern dwarf mistletoe retain their needles over the winter, whereas brooms produced by the rust fungus, *Chrysomyxa arctostaphyli* Dietel (see Section 32), lose their needles in late fall or winter.

Eastern dwarf mistletoe is a perennial, obligate, parasitic seed plant. Its aerial shoots or basal cups are the very conspicuous signs that distinguish eastern dwarf mistletoe brooms from those of spruce broom rust, caused by *C. arctostaphyli*.

Selected bibliography

Blanchard, R.D.; Tattar, T.A. 1981. Field and laboratory guide to tree pathology. Academic Press, New York, NY. 285 p.

Boyce, J.S. 1961. Forest pathology. 3rd ed. McGraw-Hill Book Co., New York, NY. 572 p.

Hawksworth, F.G. 1967. Dwarf mistletoes, *Arceuthobium* spp. Pages 31-35 *in* A.G. Davidson and A.M. Prentice, eds. Important forest insects and diseases of mutual concern to Canada, the United States and Mexico. Dep. For. Rural Dev., Ottawa, Ont. 248 p.

Ostry, M.E.; Nicholls, T.H. 1979. Eastern dwarf mistletoe on black spruce. U.S. Dep. Agric., For. Serv. For. Insect Dis. Leaflet 158. 7 p.

Prepared by Pritam Singh.

A

B

Plate 56

A. Witches'-broom on black spruce formed as a result of infection by *Arceuthobium pusillum*, the causal agent of eastern dwarf mistletoe.

B. Pistillate plant of *Arceuthobium pusillum*, the causal agent of eastern dwarf mistletoe, with nearly mature fruit. The host plant is black spruce.

57. Black knot
Apiosporina morbosa **(Schwein. : Fr.) v. Arx**
Plate 57

Hosts: Mainly cherry and plum; rarely peach.

Distribution: Common throughout eastern Canada.

Effects on hosts: This fungus produces cankers and can cause considerable branch and twig mortality. It may also induce severe stunting and death of trees. Damage due to black knot is unsightly on trees used for ornamental purposes and reduces fruit production. Black cherry is the only host reaching a commercial size in the forest, and cankers on the trunk create entry points for fungi causing wood decay.

Identifying features: Only woody parts of the trees are attacked, usually the twigs and small branches. The name "black knot" describes the black or charcoal-like, rough, spindle-shaped, cankerous overgrowth or tumors, usually occurring on one side of the infected twig or branch but often encircling it. The knots vary from 4 mm to 4 cm in diameter and from 8 mm to 20 cm in length.

The initial evidence of the disease is a slight swelling of the infected branch. As the swelling increases in size, the bark ruptures, resulting in the formation of a light brown to olive-green knot, which ultimately turns black. Those portions of the twigs or branches beyond the knots often die. As the

A

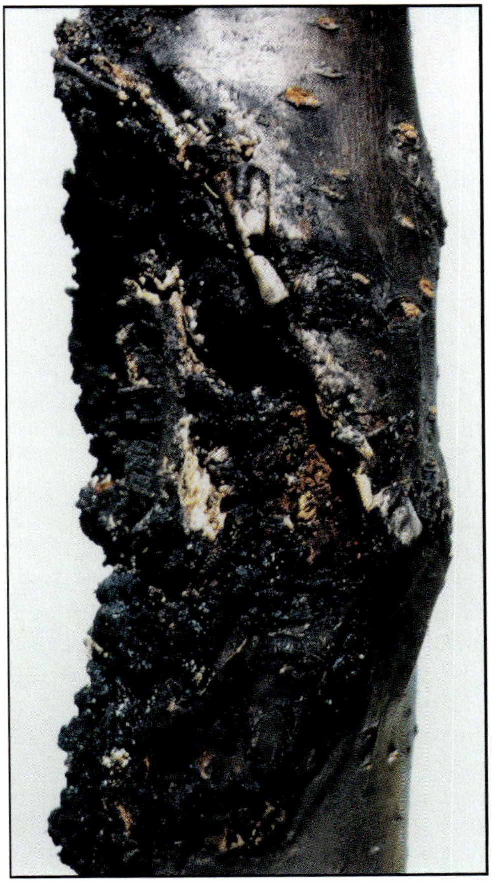

B

C

Plate 57

A. Fruiting bodies of *Apiosporina morbosa*, the cause of black knot, on cherry.

B. Black knot, caused by *Apiosporina morbosa*, causing a canker-like development on the main stem of a young cherry.

C. A typical black knot on a plum branch, caused by *Apiosporina morbosa*.

knots age, they become riddled with insect galleries.

Large amounts of gum are occasionally produced by the host in the cankered areas during humid weather. At times, a fungal parasite of *A. morbosa, Scopinella sphaerophila* (Peck) Malloch, gives a downy appearance to the knot.

Life history: The initial infection starts in the spring, but the swelling is not evident until the fall or the following spring. At this time, the bark ruptures, and a light yellow-green growth fills the crevices. In late spring or early summer, the infected area takes on a pale green tinge, followed by the formation of an olive-green velvety layer over the surface of the gall. This layer is composed of spores, which are dispersed by wind. In late summer, the velvety layer disappears, and the knots become progressively darker in color, ultimately turning black and hardening. In winter, another type of spore develops in flask-shaped structures that are formed in the knots. The following spring, these spores are discharged and dispersed to the new healthy branches, where they germinate to produce fresh infections, thus completing a 2-year life cycle. Center areas or older parts of the knots die in the second or third year of infection, but the fungus continues to grow at the periphery and spreads to the adjoining tissues until the branch is encircled, after which it spreads more rapidly along the linear axis of the branch. Old infections remain on branches for several years. Although smaller branches may be killed within a year after infection, the larger branches usually resist attack or spread of infection for several years. Most of the trunk knots originate from infections on small lateral branches.

Control: All the diseased twigs should be cut about 15 cm back of the knot in late fall. This should be done using tools swabbed with alcohol (70%) between cuts. The material should then be destroyed. Cankers on the trunks of high-value trees must be cut out so that a margin of healthy bark remains. It is recommended that all infested, wild, worthless fruit trees in the vicinity be destroyed. A fungicide can be effective in preventing infection and should be applied as soon as the buds begin to open, again when the flower buds begin to open, and when the blossom petals fall. Some host varieties possess a degree of resistance and should be selected if replanting is being contemplated. Two fungi, in addition to *S. sphaerophila*, are presumably effective as biological control agents: *Trichothecium roseum* (Hoffm.) Link : Fr., which appears in July and August; and *Coniothyrium* sp., which is frequently isolated from the knots.

Additional information: In early literature, *A. morbosa* was referred to as *Dibotryon morbosum* (Schwein. : Fr.) Theiss. & Sydow, the asexual form being a *Cladosporium*.

Selected bibliography

Corlett, M. 1976. *Apiosporina morbosa*. Agric. Can., Ottawa, Ont. Fungi canadenses 84. 2 p.

Davidson, T.R. 1973. Diseases, insects and mites of stone fruits. 3rd rev. Agric. Can., Ottawa, Ont. Publ. No. 915. 59 p.

Prepared by G. Laflamme and Pritam Singh.

58. Red belt fungus
Fomitopsis pinicola (Swartz : Fr.) P. Karsten
Plate 58

Hosts: Trembling aspen, white and yellow birch, balsam fir, eastern hemlock, ironwood, sugar maple, eastern white, jack, and red pine, and black and white spruce.

Distribution: Widespread in eastern Canada.

Effects on hosts: This fungus, often called the brown crumbly rot fungus, produces a brown rot and is capable of causing heart rot in living trees, but it is most important as a destroyer of wood in trees killed by other agents. It has been identified as the most frequently encountered brown sap rot in jack pine, red pine, and eastern white pine killed by fire in Ontario.

Identifying features: The early stage of the decay is characterized by a pale yellow to brownish color of the wood. Eventually the decay produces an obvious yellow-brown or reddish color, with the wood completely crumbling, often fracturing into cube-like sections. The fruiting bodies are formed on the trunk of the host and are either shelf-like or hoof-shaped. The upper surface in young fruiting bodies may be reddish to dark brown or almost black, with a distinct, resinous crust. As the structure ages, the older parts turn more gray and furrowed. When active, the undersurface is white to yellowish, becoming light brown when drying. A wide band of red color on the margin of the fruiting body is very distinctive and characteristic of the fungus, although it is not always present.

Life history: This fungus gains entry to its hosts through wounds or holes bored by insects. As it most often colonizes dead material, there are usually many points where infection can be initiated. Once established, the fungus causes decay and produces its perennial fruiting bodies on the external surface of successfully colonized hosts. Spores are wind-borne and are probably produced throughout the growing season when moisture is abundant.

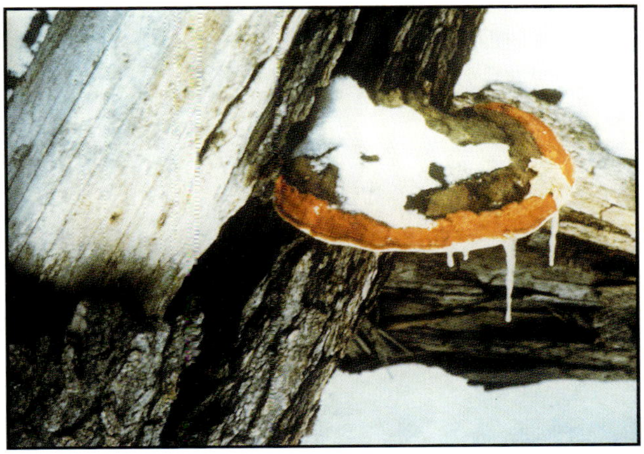

Control: Prompt salvage operations in areas where trees have been killed by fire, blow-down, insects, or fungi reduce the damage caused by this fungus.

Additional information: This fungus was known for many years as *Fomes pinicola* (Swartz : Fr.) Cooke and is found under this name in early references.

Samples for identification should include the fruiting body or stem sections with a very early stage of decay. The presence of brown cubical rot alone is not adequate for identifying this fungus, as a number of other fungi cause this type of decay.

Selected bibliography

Basham, J.T. 1957. The deterioration by fungi of jack, red and white pine killed by fire in Ontario. Can. J. Bot. 35:155-172.
Boyce, J.S. 1961. Forest pathology. 3rd ed. McGraw-Hill Book Co., New York, NY. 572 p.
Hepting, G.H. 1971. Diseases of forest and shade trees of the United States. U.S. Dep. Agric., For. Serv. Agric. Handb. No. 386. 658 p.

Prepared by D.T. Myren.

Plate 58

Fruiting bodies of *Fomitopsis pinicola*, the red belt fungus, on dead hardwood. Although capable of infecting living trees, the fungus is usually found as the cause of brown crumbly rot on dead hardwoods and conifers. (Photograph courtesy of D. Ropke.)

59. Red ring rot
Phellinus pini (Brot. : Fr.) A. Ames
Plate 59

Hosts: Mainly balsam fir, eastern white and jack pine, and black and white spruce; occasionally tamarack.

Distribution: Widely distributed in eastern Canada.

Effects on hosts: This fungus produces a white pocket rot in the trunk of its hosts, causing both a loss of volume and a decrease in the grade of milled products. It does not kill its hosts but acts as a heart rot, and it is considered one of the major causes of volume loss in conifers in North America. Red ring rot is not usually considered a butt rot, but it can cause considerable volume loss to the middle and upper parts of the butt log and can also move much higher.

Identifying features: The fruiting bodies of red ring rot are perennial structures and vary in shape from hoof-shaped to almost perfectly flat. The upper surface of the fruiting body is grayish to dark brown and often has concentric bands running parallel to the outside edge. The margin of an active fruiting body is yellowish brown and is somewhat velvet-like on the top. The lower surface of the fruiting body is poroid and ranges from grayish brown to a deeper brown. Fruiting bodies on eastern white pine are often formed at branch stubs and tend to be flat, whereas those on spruce often form as a shelf.

The early stage of decay by this fungus is sometimes called "red heart" because of the reddish color of the still-firm wood and because the fungus is usually a heart rot. In some hosts, early decay is marked by the presence of purple hues in the wood. In the more advanced stages, the small, characteristic fusiform pockets of whitish tissue are evident. These pockets are not uncommon in spruce or pine sold as economy studs.

Life history: Spores can be produced by the fruiting bodies whenever moisture and temperature conditions are suitable. Most active spore production occurs in the spring and fall. The spores are wind-borne, and entry usually takes place through dead branch stubs. This fungus can also infect weevil-killed tips of white pine. Wounds do not normally seem to make good infection sites.

Control: Control of red ring rot can be achieved by harvesting before decay becomes established or too extensive. Pathological rotation periods of 160–170 years have been established for eastern white pine in Ontario. These rotation periods, however, vary between species and geographic locations.

Additional information: This fungus was known for many years as *Fomes pini* (Thore : Fr.) Karsten and earlier as *Trametes pini* (Thore) Fr.

Samples for identification should include fruiting bodies and wood in the early stages of decay.

Selected bibliography

Boyce, J.S. 1961. Forest pathology. 3rd ed. McGraw-Hill Book Co., New York, NY. 572 p.

Hepting, G.H. 1971. Diseases of forest and shade trees of the United States. U.S. Dep. Agric., For. Serv. Agric. Handb. No. 386. 658 p.

Prepared by D.T. Myren.

Plate 59

A. Fruiting bodies of *Phellinus pini*, the red ring rot fungus, on the end of a spruce log. The fungus attacks living, dead, and down conifers, causing a white pocket rot.

B. Fruiting bodies of *Phellinus pini*, the causal agent of red ring rot, on a branch stub of a living eastern white pine. This fungus causes a white pocket rot.

C. Fruiting body of *Phellinus tremulae*, the false tinder fungus, on trembling aspen. *Phellinus tremulae* causes a white trunk rot of living and dead poplar and other hardwoods.

D. Fruiting bodies of *Ganoderma tsugae*, the lacquer conk fungus, on hemlock. The fungus causes a soft spongy white rot of hemlock, spruce, and pine. (Photograph courtesy of D. Ropke.)

E. Fruiting body of *Fomes fomentarius*, the tinder fungus, on a dead yellow birch. This fungus causes a white rot of dead hardwoods, although living trees are occasionally infected.

F. Sterile black conk of the clinker fungus, *Inonotus obliquus*, on living yellow birch. The fungus is usually found on living birch and produces a white rot.

G. *Trichaptum abietinum*, the purple conk fungus, fruiting on conifer disk. It causes a white pocket rot, often called pitted sap rot, in dead conifer timber and rarely in living trees.

H. *Ganoderma applanatum,* the artist's conk fungus, on a living American beech. *Ganoderma applanatum* causes a white mottled rot mainly on dead hardwoods, but living trees and conifers are occasionally attacked. (Photograph courtesy of C. Moffet.)

I. Fruiting bodies of *Polyporus squamosus*, the scaly polyporus fungus, on a maple stump. This fungus causes a white rot in living hardwoods.

J. A living red oak with fruiting bodies of the sulfur fungus, *Laetiporus sulphureus*. This decay fungus causes a brown cubical rot in living hardwoods, occasionally conifers, and may invade dead trees and stumps.

K. *Climacodon septentrionalis*, the northern tooth fungus, on sugar maple. The fungus causes a soft spongy white rot of heartwood of living maples and other hardwoods.

A

B

C

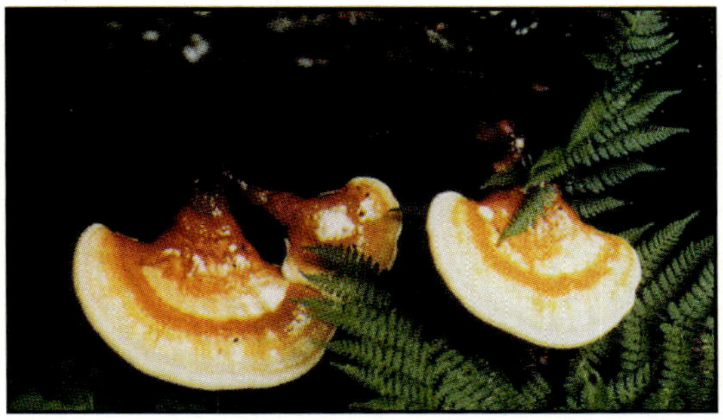

D

E

F

G

H

I

J

K

60. White spongy rot
Phellinus everhartii (Ell. & Gall.) A. Ames
Plate 60

Hosts: Red and white oak.

Distribution: Reported only from Ontario, Prince Edward Island, and Quebec.

Effects on hosts: This fungus produces a white spongy or flaky heart rot in trunks of living trees and can also attack living sapwood. The decay is usually in the lower portion of the trunk, thus reducing the volume in the butt log. *Phellinus everhartii* is considered moderate in its rate of decay.

Identifying features: The woody perennial fruiting bodies of this fungus are formed on the trunk of the host and are shelf- or hoof-shaped. They are often associated with swellings or cankers. They are brown when young but turn black and become furrowed with age. Older fruiting bodies are rough and develop very noticeable cracks. The margins remain brown. The undersurface is poroid and brown to reddish brown. Fruiting bodies can reach a width of 36 cm and project up to 15 cm from the trunk. The rot is characterized by a flaky nature, as the decay occurs most rapidly between the rays. Dark brown lines called zone lines are evident in the wood when an advanced stage of decay is reached. Burls on oak are an outward symptom of the presence of this fungus in the tree, but this is not true in all cases and cannot be relied upon as the sole proof of infection.

Life history: Entry occurs through trunk wounds. Once decay is established, production of fruiting bodies is initiated. The fruiting bodies develop on the trunk and also on felled infected trees. Spore production occurs during the growing season. The spores are wind-borne.

Control: As trunk wounds are the major infection sites, steps should be taken to minimize wounding. Particular care should be taken around final crop trees during thinning operations.

Additional information: *Phellinus everhartii* was known as *Fomes everhartii* (Ell. & Gall.) v. Schrenk & Spauld. in early literature.

The range of oak in eastern Canada is limited, but it is managed as a commercial species in some areas. No detailed decay studies have been done with oak in Canada, but a large number of wood rotters have been collected from this host during routine forest surveys. Survey records show that these fungi are not collected frequently.

Samples for identification of *P. everhartii* should include the fruiting bodies.

Selected bibliography

Boyce, J.S. 1961. Forest pathology. 3rd ed. McGraw-Hill Book Co., New York, NY. 572 p.

Hepting, G.H. 1971. Diseases of forest and shade trees of the United States. U.S. Dep. Agric., For. Serv. Agric. Handb. No. 386. 658 p.

Prepared by D.T. Myren.

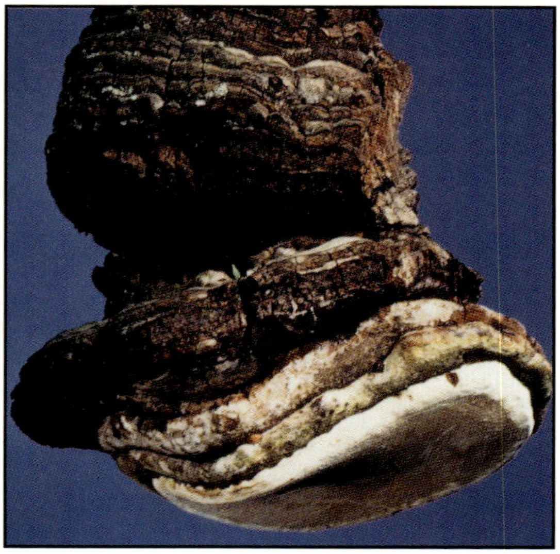

Plate 60

Fruiting body of *Phellinus everhartii*, the white spongy rot fungus. The fungus causes a soft white spongy or yellow flaky heart rot of living hardwoods, particularly oak.

61. **Clavariform juniper rust**
Gymnosporangium clavariiforme **(Wulfen : Pers.) DC.**
Plate 61

Hosts: Common juniper, with serviceberries and hawthorn as alternate hosts.

Distribution: Found in sporadic patches in all provinces of eastern Canada except Newfoundland.

Effects on hosts: In older infections of juniper, the fungus causes fusiform swellings on branches; occasionally, in severe infections, it induces witches'-brooms. However, in younger or recent infections, no symptoms are produced except for the appearance of pustules on twigs and young shoots.

On serviceberries and hawthorn, the rust causes mummification of fruits and yellow-brown spots on leaves.

Identifying features: The rust is easily identifiable on juniper by the presence of brown-orange, cylindrical pustules on fusiform branch swellings.

On serviceberries and hawthorn, the rust is seen as cinnamon-brown, ruptured pustules formed on mummified and deformed fruits and small twigs and on the undersurfaces of leaves.

Life history: One stage of the fungus matures on fruits, twigs, or leaves of serviceberries or hawthorn during midsummer and discharges minute, cinnamon-brown spores, which germinate on juniper foliage and penetrate young shoots. The fungus remains in a vegetative state during the winter and then produces orange-brown spores in cylindrical pustules on the branch swellings the following spring. Mature spores of these pustules release another type of small spore during wet periods; these spores disperse and infect foliage and fruits of serviceberries and hawthorn.

Control: Because the rust is of no significance to forestry in eastern Canada, there are no known control measures. Removal of one host from an area would result in some control. Sanitation and fungicides could be employed to protect ornamental trees.

Additional information: *Gymnosporangium clavipes* (Cooke & Peck) Cooke & Peck (see Section 62) is more common than *G. clavariiforme*, and both can be found on serviceberries at about the same time of year. They can be distinguished easily by spore color: the former has bright orange spores, and the latter has browner spores. On juniper, *G. clavipes* fruiting bodies are cushion-shaped, and the spores are orange to bright red. The fruiting bodies of *G. clavariiforme* on juniper are pillar-like, and the spores are brownish orange.

Selected bibliography
Parmelee, J.A. 1965. The genus *Gymnosporangium* in eastern Canada. Can. J. Bot. 43:239-267.

Ziller, W.G. 1974. The tree rusts of western Canada. Environ. Can., Can. For. Serv., Victoria, B.C. Publ. No. 1329. 272 p.

Prepared by Pritam Singh.

Plate 61

Gelatinous fruiting bodies of *Gymnosporangium clavariiforme*, the causal agent of clavariform juniper rust, on juniper. (Photograph courtesy of M. Dumas.)

62. **Quince rust**
Gymnosporangium clavipes **(Cooke & Peck) Cooke & Peck**
Plate 62

Hosts: Mainly common juniper; occasionally creeping juniper and eastern red cedar; alternate host mainly serviceberries, occasionally apple, mountain-ash, and hawthorn.

Distribution: Occurs sporadically in small patches throughout eastern Canada.

Effects on hosts: On juniper, quince rust causes fusiform swellings of twigs and branches, resulting in malformation of shoots. In severe infections, it causes stunting and mortality of branches or even small trees.

On serviceberries, quince rust causes mummification of infected fruits and swelling of petioles and small twigs.

Identifying features: On juniper, the pathogen is usually found on stems, but it can also occur on leaves. It produces fusiform swellings or galls on twigs, small branches, and main stems. These galls can be nodal or internodal and encircle the infected part. At maturity, they rupture irregularly with rough, black bark, exposing cushion-shaped fruiting bodies, which produce orange or bright red powdery masses of spores.

A

B

C

Plate 62

A. Torn covers of a spore mass of *Gymnosporangium clavipes*, the cause of quince rust, on fruits of hawthorn. The covers tear in this characteristic manner at the time of spore release.

B. Torn covers of a spore mass of *Gymnosporangium clavipes*, the cause of quince rust, on stems of hawthorn. The covers tear in this characteristic manner at the time of spore release.

C. Fruiting of *Gymnosporangium clavipes*, the cause of quince rust, on juniper. The fruiting bodies expand in wet weather. (Photograph courtesy of M. Dumas.)

On serviceberries, the infection is mostly on fruits and leaf petioles and rarely on leaves, stems, and branches. Mummified fruits and swollen petioles are typical symptoms.

Life history: The fungus overwinters in the inner bark of the infected juniper and produces spores in the early spring. These spores germinate, producing spores that infect the new fruit, shoots, and leaves of the alternate host. Fertilization occurs on the alternate host and spores are soon formed, which infect juniper. Spores are formed on the juniper branches the following spring to continue the life cycle. The fungus persists for several years in juniper stems, but it is annual in juniper foliage and on the alternate host.

Control: Because the disease is of no importance to forestry, no control measures are considered. However, several effective fungicidal sprays can be used to combat the disease on ornamental junipers. Pruning and burning of infected branches are also recommended for ornamentals. Removal of one of the hosts from the area of the host to be protected can also be effective. The fungal mycelium in juniper is confined to the outermost layers of the living inner bark, and infections can be eliminated by scraping off the bark about an inch around the gall.

Additional information: Often *Gymnosporangium clavariiforme* (Pers.) DC. (see Section 61) is found on the same host as *G. clavipes*, and distinguishing between the two pathogens is not too difficult with good specimens. Fruiting bodies of *G. clavariiforme* on juniper are brownish orange and cylindrical, whereas they are orangish-bright red and cushion-shaped in *G. clavipes.* More than 480 species of deciduous hosts have been identified for this rust.

Selected bibliography

Crowell, I.H. 1935. The hosts, life history and control of *Gymnosporangium clavipes* C. and P. J. Arnold Arbor. Harv. Univ. 16:367-410.

Ziller, W.G. 1974. The tree rusts of western Canada. Environ. Can., Can. For. Serv., Victoria, B.C. Publ. No. 1329. 272 p.

Prepared by Pritam Singh.

63. Comandra blister rust
Cronartium comandrae Peck
Plate 63

Hosts: Mainly jack pine; occasionally mugho, pitch, and Scots pine; alternate hosts are bastard and northern bastard toadflax.

Distribution: Found throughout the range of its hosts in eastern Canada.

Effects on hosts: Cankers caused by the rust commonly girdle and kill branches and stems in trees of all ages. Seedlings are often killed, and large trees frequently suffer branch and upper crown mortality. Pines usually produce copious amounts of resin in the cankered area, and the rust causes swelling of tissues associated with the cankers. Bark ruptures occur when the spores are released, and this contributes to the desiccation and death of cankered tissue.

Identifying features: The cankers are most conspicuous in the spring when the orange-vermillion pustules of spores are formed on the infected bark. The teardrop shape of these spores, apparent at hand-lens magnification, is diagnostic. The spindle-shaped cankers are somewhat similar to those caused by other stem rusts and, unlike those of sweet fern blister rust (see Section 64), are not confined to the base of the stem. An abundance of toadflax in the area suggests the occurrence of this rust.

Life history: This fungus has the five fruiting stages typical of many rust fungi. The orange-vermillion spores formed in the spring and early summer infect toadflax. Two fruiting stages occur on the herbaceous hosts. Early in the summer, small, dome-shaped, yellow pustules produce spores that infect other toadflax. Later, hair-like, translucent yellow structures are produced on the toadflax, and the spores on these germinate in place to form the spores that infect pines. A first fruiting stage on pine has a sexual function and occurs in the early summer. Small, inconspicuous fruiting structures of this stage form in the bark and exude droplets of spores. Spore droplets are easily washed away by rain, and this stage is not very conspicuous. The orange-vermillion spores develop the following spring or in the spring of the second or third year.

Control: Branches with cankers should be removed to prevent spread to the stem. This is practical only for high-value trees, such as ornamentals or those in seed production areas. Control in plantations and natural stands has not been necessary, because only a few isolated occurrences of extensive damage have been reported. Eradication of toadflax is difficult because new shoots arise each year from a root-like rhizome beneath the soil surface. Nurseries should not be located near concentrations of toadflax. Although no safe distance has been identified, a distance of 1 km is recommended.

Additional information: Northern bastard toadflax seems to be a herbaceous host of relatively minor importance. This plant occupies a moister site and does not seem to be as susceptible to infection as bastard toadflax, which grows on more extensive and drier sites.

Selected bibliography

Hiratsuka, Y.; Powell, J.M. 1976. Pine stem rusts of Canada. Environ. Can., Can. For. Serv. For. Tech. Rep. No. 4. 82 p.

Ziller, W.G. 1974. The tree rusts of western Canada. Environ. Can., Can. For. Serv., Victoria, B.C. Publ. No. 1329. 272 p.

Prepared by H. Gross.

Plate 63

A. Jack pine stem with mature spores of *Cronartium comandrae*, the causal agent of comandra blister rust, in which the white membrane covering the spore mass is just starting to tear open to release the spores.

B. A toadflax leaf with hair-like fruiting bodies of *Cronartium comandrae*, the causal agent of comandra blister rust.

A

B

64. **Sweet fern blister rust**
***Cronartium comptoniae* Arthur**
Plate 64

Hosts: Mainly jack pine; occasionally mugho, pitch, and Scots pine; alternate hosts are sweet fern and sweet gale.

Distribution: Found throughout the range of its hosts in eastern Canada.

Effects on hosts: This fungus commonly infects and kills pine seedlings. Presence of the fungus causes blister-like swellings of the bark and wood. Cankers can girdle the stem, and young trees often die when the fungal fruiting ruptures the bark. Rodents seem to prefer infected bark as food, and their feeding can girdle and thus kill trees. As trees age, they seem to become tolerant of the presence of the canker, possibly because of their thicker bark. However, the disease continues to cause distortion. Cankers usually extend to a height of 1 m or more and frequently serve as entry points for decay fungi.

Identifying features: The swellings or blisters that occur at the base of infected trees are the most apparent symptom of the disease on pines. In late spring, the fungus fruits on these blisters and produces orange masses of spores. A gnarled shape and blister-like overgrowths are typical of sweet fern blister rust. Two fruiting stages occur on the

A

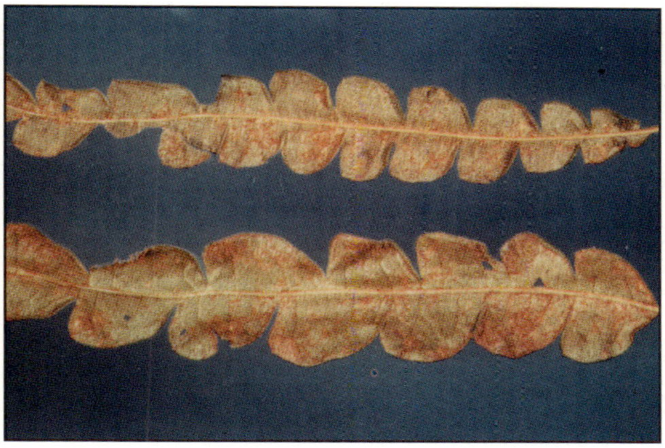

B

Plate 64

A. Stem canker on jack pine caused by *Cronartium comptoniae*, the causal agent of sweet fern blister rust.

B. Fruiting of *Cronartium comptoniae*, the causal agent of sweet fern blister rust, on sweet fern leaves.

foliage of the herbaceous hosts: one forms small yellow pustules about the size of a pinhead starting about midsummer, and the other produces reddish-brown, hair-like structures later in the summer. Both stages are formed on the lower leaf surface.

Life history: This fungus has the five spore stages typical of many rust fungi. Spores produced in late summer on the herbaceous host foliage germinate to produce the spores that infect pine. In the fall, another stage, which has a sexual function, occurs on the pines. In late spring, the orange spores, which are the most typical fruiting stage, develop and are released to infect the herbaceous hosts. On these hosts, spores produced in early summer have several generations and spread the infection to other sweet fern and sweet gale. Later in the summer, the hair-like structure develops and germinates to produce the spores that infect the pine and thus complete the cycle.

Infection can occur on 1-year-old seedlings; with most pines, most of the infection occurs within the tree's first 4 years. Once the tree reaches a basal diameter of about 7 cm, it is very unlikely to become infected. The orange fruiting stage is usually confined to the blister-like overgrowths, and some fruiting seems to occur for most years throughout the lives of the canker and tree.

Control: Older seedlings should be used as planting stock where a high rust hazard exists with respect to the abundance of the herbaceous host. This limits the use of most containerized planting stock.

Because swellings caused by infection by stem rusts and gall rusts are evident usually within a year after infection, nursery inspections and grading rules should function to ensure that infected trees are destroyed.

Studies indicate that hazard diminishes rapidly beyond a distance of 15 m from sweet fern. Hence, nurseries should be located at a reasonable distance from concentrations of the herbaceous hosts. A distance of 0.5 km should be satisfactory. At existing nurseries, pine host seedlings should be grown at locations as far from herbaceous hosts as possible or where the herbaceous hosts have been eradicated from a zone at least 30 m in width.

Additional information: Sweet fern and jack pine are well adapted to dry, sandy sites, and extensive stands containing both hosts are common. Sweet gale requires moist sites and often occurs along lakeshores and stream banks. Other similar stem rusts are orange stalactiform blister rust (*Cronartium coleosporioides* Arth. f. sp. *coleosporioides*) and comandra blister rust (see Section 63). Cankers produced by these rusts are not typically confined to the base of the stem. Comandra blister rust produces spore masses that have a deeper orange color shaded more towards vermillion, and the spores have tail-shaped protuberances that are apparent at hand-lens magnification. Rodents vigorously seek out bark infested by orange stalactiform blister rust, and cankers often show exposed wood and have the typically elongate shape.

Selected bibliography

Hiratsuka, Y.; Powell, J.M. 1976. Pine stem rusts of Canada. Environ. Can., Can. For. Serv. For. Tech. Rep. No. 4. 82 p.

Ziller, W.G. 1974. The tree rusts of western Canada. Environ. Can., Can. For. Serv., Victoria, B.C. Publ. No. 1329. 272 p.

Prepared by H. Gross.

---

65. Eastern gall rust
Cronartium quercuum (Berk.) Miyabe ex Shirai
Plate 65

Hosts: Mainly jack and Scots pine; occasionally Austrian pine; alternate hosts are red and occasionally other oaks.

Distribution: In eastern Canada, known only from central and southern Ontario.

Effects on hosts: Galls produced as a result of infection weaken the stem, making the host more susceptible to windbreakage. Portions of infected branches or twigs distal to the gall may die. The disease has caused some problems in nurseries, resulting in cull of pine seedlings. On oaks, the disease causes only minor leaf spotting and is not considered of any significance.

Identifying features: The development of globose galls on the pine hosts is characteristic of this rust. In the spring, spore production on the galls occurs under a white, blister-like structure that soon bursts,

A

B

C

Plate 65

A. Fruiting of *Cronartium quercuum*, the eastern gall rust fungus, on a typical globose gall on jack pine.

B. Hair-like fruiting of *Cronartium quercuum*, the eastern gall rust fungus, on the lower surface of a red oak leaf.

C. A close-up of the hair-like fruiting of *Cronartium quercuum*, the eastern gall rust fungus, on the lower surface of a red oak leaf.

releasing the bright orange spores. On oak, small brown spots form on the upper leaf surface in the spring, and small yellowish spore pustules develop on the lower surface directly below them. In the late spring or early summer, the spore pustules are replaced by brown, bristle-like structures, which remain for the rest of the year.

Life history: Orange spores are produced on the gall on the pines in the spring and infect the young oak leaves. Yellow spore pustules develop on the infected leaves and produce spores that infect other oaks. In the early summer, these pustules are replaced by bristle-like structures that germinate in place and produce the spores that infect the pine. About 2 years after infection, orange droplets form on the newly developing gall. These droplets contain spermatia, which are involved in fertilization. The following year, white blisters form on the gall beneath which orange spores develop. The blisters soon rupture, and the spores are released to infect the oak, repeating the cycle.

Control: Infection in nurseries can be reduced by using fungicides to protect the pine or by removing the alternate host. No control measures have been attempted in forested areas.

Additional information: A globose gall on pine stems is also caused by *Endocronartium harknessii*

(J.P. Moore) Y. Hirats. (see Section 66), which can go directly from pine to pine without an alternate host. In areas where oaks are present, examination of germinating spores with a microscope is necessary to separate *C. quercuum* from *E. harknessii*. Recently, four *formae speciales* have been erected for *C. quercuum*. *Cronartium quercuum* (Berk.) Miyabe ex Shirai f. sp. *banksianae* Burdsull and G. Snow, which attacks jack pine, is the only one of importance to eastern Canada.

Selected bibliography

Hepting, G.H. 1971. Diseases of forest and shade trees of the United States. U.S. Dep. Agric., For. Serv. Agric. Handb. No. 386. 658 p.

Hiratsuka, Y.; Powell, J.M. 1976. Pine stem rusts of Canada. Environ. Can., Can. For. Serv. For. Tech. Rep. No. 4. 82 p.

Prepared by D.T. Myren.

66. Western gall rust
Endocronartium harknessii (J.P. Moore) Y. Hirats.
Plate 66

Hosts: Mainly jack and Scots pine; occasionally Austrian and mugho pine.

Distribution: Found throughout the range of its hosts in eastern Canada.

Effects on hosts: Galls caused by this rust fungus usually encircle the affected pine stem or branch. White blisters under which orange spores develop erupt through the bark in the spring, and the associated tissue desiccation often kills the galled area. Tree parts distal to the dead galls subsequently die, and young trees are frequently killed. Rodents feed on rust galls during the winter, and this also causes considerable mortality in some years.

Galls distort the growth form and shape of associated tree parts, and large numbers of galls on a tree detract from the aesthetic appearance. This is especially important with respect to ornamentals and Christmas trees.

Identifying features: The globose galls caused by the disease are the most distinguishing feature. These are most conspicuous in the spring, when they are covered with bright orange spores.

Life history: This fungus does not have an alternate host and therefore has a shortened life cycle. The orange spores produced on the galls in the spring cause direct infection of other pines.

Control: Galls on branches can be removed by pruning, and shearing Christmas trees provides some control. Fungicides applied during the period when spores are abundant can provide disease control for nurseries and ornamental trees. Galled trees in the vicinity of nurseries should be removed.

Additional information: Eastern gall rust (see Section 65) causes galls that appear identical to galls caused by the western gall rust. When the alternate host for eastern gall rust—oak—is present, galls could be the result of infection by either of these fungi. Germination and examinination of the spores under a microscope provide the best means of differentiating these rusts under those circumstances.

Selected bibliography

Hiratsuka, Y.; Powell, J.M. 1976. Pine stem rusts of Canada. Environ. Can., Can. For. Serv. For. Tech. Rep. No. 4. 82 p.

Ziller, W.G. 1974. The tree rusts of western Canada. Environ. Can., Can. For. Serv., Victoria, B.C. Publ. No. 1329. 272 p.

Prepared by H.L. Gross and D.T. Myren.

Plate 66

A. Fruiting of *Endocronartium harknessii*, the cause of western gall rust, on jack pine.

B. Jack pine seedling with a stem gall caused by *Endocronartium harknessii*, the causal agent of western gall rust.

A

B

67. **White pine blister rust**
Cronartium ribicola **J.C. Fischer**
Plate 67

Hosts: Mainly eastern white pine; alternate hosts are domestic and wild currant and gooseberry.

Distribution: Widespread in eastern Canada.

Effects on hosts: The fungus invades and kills the cells of the inner bark and recently formed woody tissue, eventually causing death of part or all of the host. Smaller pines are killed quickly. In larger trees, trunk cankers girdle the host, retard its growth, and weaken the stem. Tops of trees may break at the point of girdling, but only that part of the host distal to the canker is killed. Branch mortality, although not fatal to the trees, reduces their market value. The disease affects trees of all ages and sizes; in pole-size stands, it results in undesirable thinning.

The disease is relatively harmless on ribes bushes; leaves are cast prematurely when infection is heavy, and fruit production may be reduced.

Identifying features: On pine, early stages of infection are characterized by patches of yellow-orange bark followed by the development of spindle-shaped cankers or swellings of the infected trunk or branch. Later, the fungus produces white blisters on the canker that cover an orange spore mass. The blister soon ruptures, exposing the mature spores. Cankered bark, chewed by rodents and with abundant resinosis, and cracked bark above the canker are good indications of the presence of the disease. When a branch or trunk of a pine tree has been girdled, foliage beyond the canker becomes yellow and later red. This condition is known as "flagging" and may be evident throughout the summer and often the following season.

On ribes, yellow blisters on the undersides of leaves are most conspicuous during cool and wet weather. In mid- to late summer, short, brown, bristle-like structures develop on the undersides of the infected leaves either from yellow blisters or from separate pustules.

Life history: The blister rust fungus needs two different hosts to complete its life cycle.

On pine, infection occurs in late summer through needles, and early symptoms appear as small, yellowish spots. During the next 12–18 months the fungus moves down into the branch, causing yellowish to orange discoloration of the bark. The discoloration spreads as the fungus grows, and the bark becomes brown and swollen, producing a spindle-shaped canker. Eventually the branch or trunk is girdled, and everything above the canker dies. In the late summer of the third season (24 months after infection), honey-colored to brownish drops of liquid exude from the branch swelling. The following spring, white blisters develop on the canker and cover an orange spore mass. After the blisters break, the spores are dispersed by wind, and the bark in the area of the blister begins to darken in color and roughen in texture. These spores infect ribes leaves.

On ribes, the rust appears in the spring on the undersides of leaves as small, light-colored dots. Within a few days, these dots develop into yellowish to light orange, spore-bearing pustules. Spores produced in these pustules can infect only other ribes plants. Several generations of these spores are produced in one season, and through these the rust intensifies and spreads from ribes to ribes. In the late summer or early fall, another type of spore is produced as short, brownish, horn- or bristle-like structures on the underside of the leaf. The bristles or horns can be so numerous that they form a brownish mat. These spores germinate in place and produce another form of spore, which is dispersed by wind and can infect only pine needles.

Control: Blister rust is virtually impossible to control in large natural stands. However, in small, high-value stands or in individual ornamental trees, the disease can be controlled by pruning and burning the diseased branches or severely diseased trees; removing and destroying all ribes bushes growing within 300 m of white pine; using resistant varieties of white pine; and avoiding planting white pine where the alternate hosts are abundant.

Additional information: Samples for identification should include a pine stem with spores or well-pressed leaves of the alternate host with spore-producing structures. Cankers from the pine host without spores can sometimes be identified, but identification is time-consuming.

Selected bibliography

Benedict, W.V. 1967. White pine blister rust. Pages 185-188 *in* A.G. Davidson and R.M. Prentice, eds. Important forest insects and diseases of mutual concern to Canada, the United States and Mexico. Dep. For. Rural Dev., Ottawa, Ont. 248 p.

Boyce, J.S. 1961. Forest pathology. 3rd ed. McGraw-Hill Book Co., New York, NY. 572 p.

Nicholls, T.H.; Anderson, R.L. 1977. How to identify white pine blister rust and remove cankers. U.S. Dep. Agric., For. Serv., North Central For. Exp. Stn. 8 p.

Prepared by Pritam Singh.

Plate 67

A. An eastern white pine in which the upper portion has been killed by *Cronartium ribicola*, the causal agent of white pine blister rust.

B. Fruiting of *Cronartium ribicola*, the cause of white pine blister rust, on the main stem of an eastern white pine. The white membrane over the spore mass is just beginning to rupture.

C. A currant leaf with infection on the upper surface by *Cronartium ribicola*, the cause of white pine blister rust.

D. Infection by *Cronartium ribicola*, the cause of white pine blister rust, on the lower surface of a currant leaf, with the repeating spore stage.

E. The hair-like fruiting structures of *Cronartium ribicola*, the causal agent of white pine blister rust, which produce the spores that infect the pine host.

A

B

C

D

E

68. Balsam fir tip blight
Delphinella balsameae (Waterman) E. Müller
Plate 68

Host: Balsam fir.

Distribution: Common in scattered patches throughout the range of its host in eastern Canada.

Effects on host: The fungus kills the needles and the current year's shoots. Although only a few branches per tree are damaged, the presence of the disease in the mid-crown lowers the value of the host as an ornamental or Christmas tree. The disease is known to affect trees of all ages.

Identifying features: Needles on the current year's shoots are killed and shrivel and curl in a characteristic fashion. The outer portions of branches and twigs become red and die. A constriction usually develops at the junction of living and dead portions of the branches. The dead needles are cast in the late fall or early spring, leaving the shoots bare.

Tiny black fruiting bodies on the upper surfaces of the dead needles distinguish the damage from similar symptoms produced by sawyer beetles, hailstones, and fusicoccum canker.

Life history: Fruiting bodies develop on the infected needles and shoots shortly after they are killed. They mature very slowly and cast their spores the following spring. The newly developing foliage is infected, and the cycle is repeated.

Control: Because the disease is not significantly damaging in forests, no control is recommended. However, pruning and destruction of affected twigs or branches back to a point where the inner bark is green are suggested for ornamental and shade trees and for high-value plantations.

Additional information: Symptoms can be distinguished from those produced by hailstones, sawyer beetles, and fusicoccum canker by laboratory examination or by needle symptoms if observed early in disease development.

Delphinella balsameae used to be known as *Rehmiellopsis balsameae* Waterman.

Selected bibliography
Boyce, J.S. 1961. Forest pathology. 3rd ed. McGraw-Hill Book Co., New York, NY. 572 p.
Smith, C.C.; Newell, W.R.; Renault, T.R. 1981. Common insects and diseases of balsam fir Christmas trees. Environ. Can., Can. For. Serv. Publ. No. 1328. 60 p.
Waterman, A.M. 1945. Tip blight of species of *Abies* caused by a new species of *Rehmiellopsis*. J. Agric. Res. 70:315-337.

Prepared by Pritam Singh.

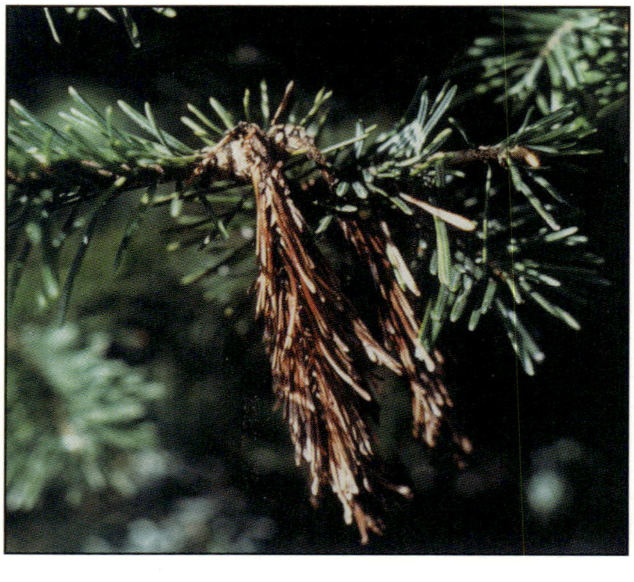

Plate 68

Balsam fir shoots infected by *Delphinella balsameae*, the cause of balsam fir tip blight.

69. Red flag of balsam fir
Fusicoccum abietinum (R. Hartig) Prill. & Delacr.
Plate 69

Host: Balsam fir.

Distribution: Found in New Brunswick, Nova Scotia, and Quebec.

Effects on host: Usually only a few red flags are found on affected trees. These become important when the appearance of the tree is affected, such as on Christmas trees or ornamentals. Christmas trees may be culled because of red flags.

Identifying features: The tips of infected twigs or branches become red and die. A constriction develops at the junction of green and red portions of the affected branch, and there may be a group of very small black pustules in this sunken area.

Life history: Spores produced from tiny black fruiting bodies cause infection. Soon after the branch is girdled, a sunken canker is formed, and the discoloration of the distal end of the branch begins. Red flags are visible from early summer until late fall or even the following spring when red needles fall, leaving the branch tip bare.

Control: Pruning the red flag several centimeters below the constriction into the green portion eliminates the symptomatic portion and reduces the spore load for further infection.

Additional information: Red flags are also caused by the feeding of adult sawyer beetles, but the feeding scar, mostly on the underside of the twig, should help in differentiation. Hail damage on the upper side of the branch and other types of mechanical injury to branches also result in red flags. Samples sent for diagnosis must contain the sunken canker and at least 5 cm of the branch above and, more importantly, below the constriction.

Selected reference

Smith, C.C.; Newell, W.R.; Renault, T.R. 1981. Common insects and diseases of balsam fir Christmas trees. Environ. Can., Can. For. Serv. Publ. No. 1328. 60 p.

Prepared by L.P. Magasi.

Plate 69

A. Red flag on balsam fir caused by *Fusicoccum abietinum.*

B. Balsam fir branch infected by *Fusicoccum abietinum*, showing the resulting red flag.

A

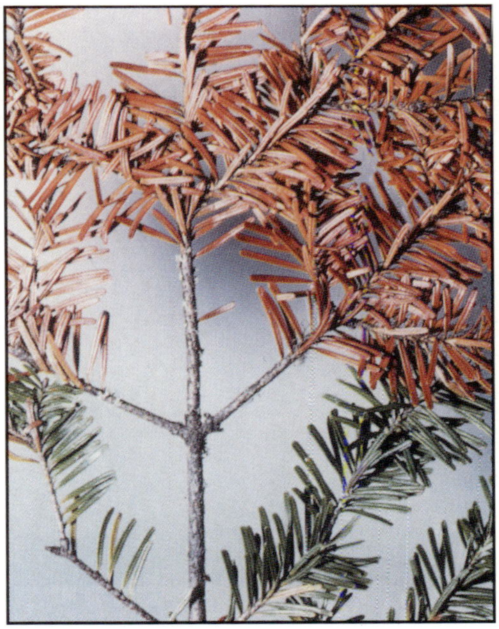
B

70. **Smothering disease** ***Thelephora terrestris*** **Ehrh. : Fr.** Plate 70

Hosts: Mainly coniferous seedlings; occasionally deciduous seedlings.

Distribution: Widespread in eastern Canada.

Effects on hosts: The relatively large fruiting bodies grow rapidly around the lower stem of seedlings and smother them without any infection of the tissue. Small seedlings may be completely covered by the fungus. Usually only a small percentage of seedlings is affected. Profuse growth of the fungus on container stock has prevented water from reaching the growing medium and resulted in mortality.

Identifying features: The fruiting body is large, up to 15 cm in diameter, producing many imbricating rosettes and surrounding the stem at the base of the seedling. It is dark brown to rusty in color, leathery, and its upper surface is kilting and hairy.

Life history: This fungus grows in humus as a saprophyte; it has also been known to occur on pine in mycorrhizal association. That a fungus could be beneficial to a plant and also smother it seems contradictory, but the smothering is an accident, as the fungus simply finds a support upon which to develop its fruiting body. The fungus does not penetrate living host tissue unless it is to function in the mycorrhizal role.

Control: Live seedlings supporting this fungus can be transplanted after the fungal structures have been removed. Because the development of the fruiting structure is favored by a dense seedling growth, increasing the distance between seedlings should help reduce the damage. It has also been observed that this fungus does not affect fast-growing seedlings. Physical removal is required when the fungus interferes with watering of container stock.

Additional information: *Thelephora terrestris* is the only species of this genus that has been found to smother conifer seedlings in eastern Canada.

Selected bibliography

Boyce, J.S. 1961. Forest pathology. 3rd ed. McGraw-Hill Book Co., New York, NY. 572 p.

Weir, J.R. 1921. *Thelephora terrestris, T. fimbriata* and *T. caryophyllea* on forest tree seedlings. Phytopathology 11:141-144.

Prepared by G. Laflamme.

Plate 70

A. *Thelephora terrestris*, the cause of smothering disease, fruiting on the base of a conifer seedling as it appears in the field.

B. *Thelephora terrestris*, the cause of smothering disease, on a conifer seedling, showing detail of the fruiting structure.

A

B

71. Dutch elm disease
Ophiostoma ulmi (Buisman) Nannf.
Plate 71

Hosts: English, rock, Scotch, Siberian, slippery, and white elm and elm hybrids.

Distribution: Widespread in eastern Canada with the exception of Prince Edward Island, where it has been found only in a limited area in the western part of the province, and Newfoundland, where it is unknown. (Newfoundland is outside the geographic range of the elms native to Canada.)

Effects on hosts: The disease is a vascular wilt and results in the death of its host. It has severely reduced the elm population of eastern Canada.

Identifying features: Early infection becomes evident about midsummer, when the fungus causes a wilting and curling of leaves, which later shrivel and turn brown. Infection occurring later in the summer causes leaves to yellow and then wilt. Premature loss of affected leaves is common. Trees infected still later in the season produce smaller leaves in all or part of the crown the following spring, and small dead branches may be evident. A cross section of an infected branch often shows a brown vascular discoloration in the form of a solid ring or a series of dots. Brown streaking can be seen if the bark is peeled back to expose the wood. Bark beetle activity may also be evident.

Life history: Dutch elm disease is spread primarily by two beetles, the native and the European elm bark beetles, which lay their eggs in chambers under the bark of dead and dying elm. The fungus develops in these chambers as it grows through the inner bark area and produces spores that are carried by the newly emerging adults. The beetles feed on healthy trees, and the fungus is introduced into the tree through the feeding wounds. Once the spores are in the host, they are spread rapidly through the water conducting system and grow in a yeast-like manner, producing the toxin that causes the wilting and eventual death of the tree. The fungus can also be spread on contaminated pruning tools or through root grafts where trees grow close together.

Control: Chemicals that can be injected through the root system of individual elm trees provide good control of this disease. The cost is modest in most situations, and some high-value ornamentals can be protected for several years with this control procedure. Sanitation is probably the most practical control method and can significantly slow the advance of the fungus through an area. All dead and dying elms as well as dead limbs in otherwise healthy trees must be removed and destroyed. Bark must be removed from any material to be held for some later use. Maintenance of healthy trees

Plate 71

A. Wilt and branch dieback caused by *Ophiostoma ulmi*, the causal agent of Dutch elm disease, on a white elm.

B. Brown vascular staining caused by *Ophiostoma ulmi* in a white elm with Dutch elm disease.

A

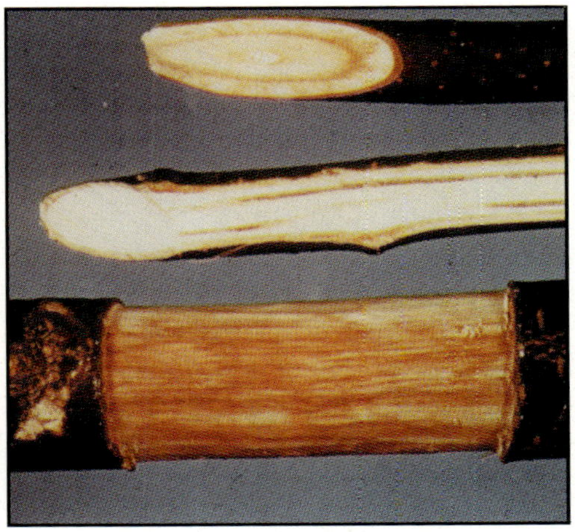

B

through proper tending and pruning is an integral part of a control program. All pruning or cutting tools used must be sterilized with alcohol between cuts when used on infected trees. Wiping the blades thoroughly with denatured alcohol between cuts on the same tree and soaking for several minutes between cuts on different trees should be satisfactory. Root grafts may need to be interrupted, but this, along with the removal of large trees, should be done by a professional arborist.

Additional information: *Pesotum ulmi* (Schwartz) Crane & Schoknecht is the imperfect state of *O. ulmi*. It will be found in early literature as *Graphium ulmi* Schwartz and more recently as *Ceratocystis ulmi* (Buisman) C. Moreau. Because the bark beetles survive in infected elm logs, felled trees containing a beetle population should not be used for fuel unless the bark has been removed.

Diagnosis of Dutch elm disease should be confirmed by culture at a laboratory. Living branches with recently wilted leaves and brown streaking under the bark or evidence of vascular discoloration in a cross section make a suitable sample.

Selected bibliography

Denyer, W.B.G. 1976. Dutch elm disease, how to recognize it, what to do about it. Environ. Can., Ottawa, Ont. Leaflet. 4 p.

Kondo, E.S.; Hiratsuka, Y.; Denyer, W.B.G., eds. 1981. Proceedings of the Dutch elm disease symposium and workshop. Environ. Can./Man. Dep. Nat. Resour., Winnipeg, Man. 517 p.

Van Sickle, G.A.; Sterner, T.E. 1976. Sanitation: a practical protection against Dutch elm disease in Fredericton, New Brunswick. Plant Dis. Rep. 60:336-338.

Prepared by D.T. Myren.

72. Verticillium wilt
Verticillium albo-atrum Reinke & Berth.
Plate 72

Hosts: Mainly maples; occasionally catalpa, sour and sweet cherry, peach, and Canada plum.

Distribution: Reported from New Brunswick, southern Ontario, Prince Edward Island, and Quebec.

Effects on hosts: This fungus causes a wilt of branches and eventually the entire tree, resulting in its death. The disease is not always fatal, but many trees die as a result of infection by this organism. The disease is found only on ornamental trees and a wide variety of herbaceous plants.

Identifying features: The disease usually appears in midsummer and is characterized by a wilt that is quite sudden. A single branch or two or the entire crown may exhibit the wilt symptoms. In maple, greenish streaks usually appear in the outer sapwood of infected branches; however, their absence does not necessarily mean that the fungus is not present. In cross sections of infected branches, these streaks appear as a solid or dotted green stain in the most recent growth rings. Other tree species may exhibit streaks of a different color, but the pattern is the same. In catalpa, for example, the discoloration is at first purplish pink, later changing to bluish brown. Many other hosts will have a brown discoloration.

Life history: The fungus is a common soil inhabitant and is widely distributed. It infects trees through wounds in the roots and spreads upwards through the vascular system, causing wilt and discoloration. Water-conducting vessels are plugged by substances produced by the fungus and by the tree in response to infection. There is evidence that toxins may be an important factor in the production of the wilt and discoloration. The fungus can be spread by the movement of soil between areas.

Control: Watering and fertilizing to maintain tree vigor are helpful. Dead branches should be removed to improve tree appearance, but currently wilting branches should not be removed until the following year, because these do recover at times. Removal of branches does not remove the fungus, as it is in the roots and trunk. Systemic fungicides hold some promise for control, but they require further research. Trees killed by verticillium wilt should be replaced by trees resistant to the disease. Conifers are immune, and suitable resistant or immune hardwoods could be suggested by a nursery.

Additional information: A similar species, *Verticillium dahliae* Kleb., also causes verticillium wilt and has been implicated in this disease in many of the hosts that can be attacked by *V. albo-atrum*. These two fungi are indistinguishable in most respects and can be separated only by laboratory studies. Branches 2–3 cm in diameter and 10–15 cm long with obvious stain make good samples.

Selected bibliography

Carter, J.C. 1975. Diseases of midwest trees. Univ. Ill. Coll. Agric./Ill. Nat. Hist. Serv. Spec. Publ. No. 35. 168 p.

Smith, L.D. 1979. Verticillium wilt of landscape trees. J. Arbor. 5:193-197.

Prepared by D.T. Myren.

Plate 72

A. Verticillium wilt, caused by *Verticillium albo-atrum*, on an ornamental maple. Note dead branches and branches with current wilt. (Photograph courtesy of R.E. Rice.)

B. Green vascular discoloration caused by *Verticillium albo-atrum*, the causal agent of verticillium wilt, in branches of an infected maple.

A

B

73. Sirococcus shoot blight
Sirococcus conigenus (DC.) P. Cannon & Minter
Plate 73

Hosts: Mainly red pine and black and white spruce; rarely jack pine and blue, Norway, and red spruce.

Distribution: Widely distributed on red pine and on the cones of white spruce in the Maritime provinces and on red pine in northwestern and central Ontario. Common on black spruce in the Laurentides Provincial Park in Quebec and has been found on spruce in a nursery in Prince Edward Island.

Effects on hosts: The disease kills only the current year's shoots. Repeated attacks have a cumulative effect, resulting first in stunted growth, then the tree succumbing to the disease. Seedlings, especially container-grown seedlings, die quickly; older, larger trees die after several successive years of severe attack.

Sirococcus shoot blight is the most serious disease in red pine plantations in Nova Scotia. In a survey in 1983, 37% of the plantations were found infected; the average frequency of infection in these was 76%, including more than 23% tree mortality.

Identifying features: On pine, needles on infected shoots wilt, collapse at the base, and bend sharply downward, giving the shoot a drooped appearance. Needles may stay on the tree for up to 2 years and undergo color changes from reddish to finally bleached straw brown. Small black fruiting bodies form at the base of infected needles, often only under the bundle sheath. Infected shoots may appear anywhere on affected trees, although lower branch infection, not to be confused with natural shading, is more common.

On spruce, the entire shoot droops, and damage appears similar to that caused by late frost.

Life history: Spores ooze out of the small black fruiting bodies in wet periods during the growing season and are carried by wind or splashed to healthy shoots by rain. Infected shoots die within 4–6 weeks. Fruiting bodies develop on newly killed needles, on cone scales, or occasionally on the dead shoot tips. The spread of the disease is usually slow, but it can intensify rapidly on infected trees. Young trees under or near infected large overstory trees are very vulnerable, as are trees in plantations with intermingling crowns. There is evidence that the fungus is seed-borne in spruce, which creates special problems for nurseries.

Control: Removal and destruction of infected shoots on ornamental trees as soon as practical, but not later than bud break, are recommended. Removal of old overstory trees in and around nurseries, plantations, and new regeneration, preferably coupled with pruning, should reduce new infections. Fungicide treatment is practical only in nurseries.

Additional information: In early literature, *S. conigenus* was first known as *Ascochyta piniperda* Lindau, then as *Sirococcus strobilinus* Preuss.

Samples should include the entire dead shoot. Fruiting bodies are found more easily in the late spring and fall.

Sclerophoma pythiophila (Corda) Höhnel is a common saprophyte on the dead needles.

Selected bibliography

O'Brien, J.T. 1973. Sirococcus shoot blight of red pine. Plant Dis. Rep. 57:246-247.

Wall, R.E.; Magasi, L.P. 1976. Environmental factors affecting sirococcus shoot blight of black spruce. Can. J. For. Res. 6:448-452.

Prepared by L.P. Magasi.

A

B

C

Plate 73

A. Red pine plantation damaged by *Sirococcus conigenus*, the causal agent of sirococcus shoot blight.

B. Red pine seedling showing the needle droop characteristic of sirococcus shoot blight caused by *Sirococcus conigenus*.

C. *Sirococcus conigenus*, the causal agent of sirococcus shoot blight, fruiting on white spruce. On pine, fruiting is often seen only after the bundle sheath is removed.

74. **Armillaria root rot**
***Armillaria mellea* complex**
Plate 74

Hosts: Mainly aspen, balsam fir, maple, oak, and jack, red, and eastern white pine, and black and white spruce. Many other hosts known.

Distribution: Common in plantations and in naturally regenerated areas throughout the forests of eastern Canada. Also one of the more important diseases in orchards, gardens, parks, campsites, and urban areas.

Effects on hosts: The fungus kills trees by girdling them at the root collar or by killing major roots. It kills the cambium and outer layers of wood and causes decay of the sapwood and heartwood of the infected roots and root collar. The decay usually does not progress upward in the stem for more than a meter. Infected trees are subject to windthrow because of weakened root systems. The fungus attacks trees of all vigor classes but is more common on trees damaged by other agents. In young trees, the disease progresses rapidly; in older trees, it progresses slowly and may never cause mortality.

Identifying features: The first evidence of the disease is a decline in vigor of the tree, progressive yellowing or browning, thinning of the crown, and eventual defoliation. This is followed by resin exudation, the formation of a canker at the point of infection on the root or at the base of the stem, and ultimate death of the tree. Often, however, the death is sudden, and many of the typical symptoms may not be observed. Soil around the roots and root collar of infected and dead trees is often resin-soaked and adheres tightly to the woody tissue.

Infected trees may occur singly or in groups. In plantations, infected trees are frequently found close to older or decaying infected stumps.

Important signs of the fungus are white or cream-colored, fan-shaped mycelial growths on the wood under the bark; dark brown to black, "shoe-string"-like structures, the rhizomorphs, on the bark of the roots and in the surrounding soil; and clusters of honey-colored mushrooms that develop on or around the base of infected trees in early autumn. The caps of these mushrooms are 5.0–12.5 cm in diameter and have dark scales on the upper surface and loosely spaced, yellowish-white gills on the lower surface. The stem of the mushroom is encircled near the cap by a thin, membranous ring.

The decayed wood is at first light brown, with a water-soaked appearance, but later it turns yellowish or whitish and attains a spongy texture in hardwoods and a stringy texture in softwoods.

Life history: The fungus is soil-borne and usually lives on dead organic material, including stumps and roots of dead trees, although it is capable of attacking living trees. The disease starts with the invasion of healthy roots by the fungal rhizomorphs that grow through soil from infected material (roots/stumps) and through root grafts between infected and healthy roots. Once established, the fungus moves up from the roots to the root collar and then to the stem, eventually killing the tree.

The mycelial fans and the rhizomorphs are found throughout the year. The mushrooms occur only during early autumn and produce spores that are dispersed by wind and may initiate infection of dead trees and stumps.

Control: Armillaria root rot is difficult to control; however, some practical measures are recommended. Careful site selection, good management, and use of trees less susceptible to infection are measures that can be used to reduce the impact of the disease. Removal of old stumps in areas to be planted is beneficial but costly. This should be considered if an area is being planted as a seed orchard or for Christmas tree production.

Additional information: What has been called *A. mellea* is now known to be a group of species and strains. These differ in pathogenicity and host preference and are the subject of much research at this time.

Armillaria ostoyae (Romagn.) Herink is the species most commonly associated with conifers in Ontario and Quebec.

Selected bibliography
Blanchard, R.D.; Tattar, T.A. 1981. Field and laboratory guide to tree pathology. Academic Press, New York, NY. 285 p.
Hepting, G.H. 1971. Diseases of forest and shade trees of the United States. U.S. Dep. Agric., For. Serv. Agric. Handb. No. 386. 658 p.

Prepared by Pritam Singh.

A

B

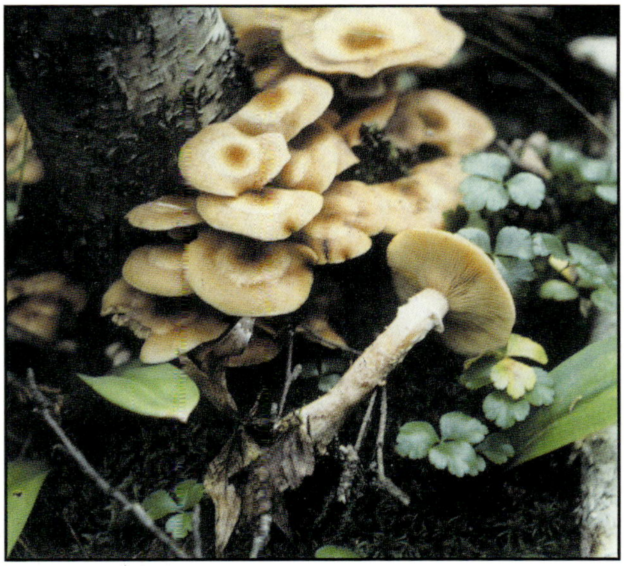

C

Plate 74

A. Mycelial fan under the bark of a white spruce infected by *Armillaria mellea*, the cause of armillaria root rot.

B. Black, strand-like rhizomorphs on the surface of a conifer root infected by the root rot fungus *Armillaria mellea*.

C. Fruiting bodies of *Armillaria mellea*, the causal agent of armillaria root rot, at the base of an infected birch.

75. Fomes root rot
Heterobasidion annosum (Fr. : Fr.) Bref.
Plate 75

Hosts: Mainly eastern white, jack, red, and Scots pine; occasionally eastern red cedar, larch, and blue spruce; rarely largetooth aspen and white elm.

Distribution: In eastern Canada, found in southern Ontario and southern Quebec.

Effects on hosts: This fungus causes root rot, resulting in the rapid death of seedlings and larger trees. Very large trees are also killed, but they usually exhibit crown symptoms and reduced growth for several years before dying. Exposed trees with root decay are prone to windthrow. Decay can extend into the lower stem and can result in significant volume loss.

Identifying features: The presence of roughly circular patches of dead trees in a conifer plantation thinned several years previously is characteristic of fomes root rot. Fruiting bodies of the fungus can be found at the base of stumps and dead or dying trees within the infected area and provide confirmation of the presence of the disease. The fruiting body is usually concealed by the litter layer, which must be pulled away from the base of the stump or tree being examined. The upper surface of the fruiting body is brown, and the undersurface is white and covered with small pores. Old fruiting bodies may be totally brown, but the poroid nature is often still evident. Needles, twigs, and other litter may be incorporated into the fruiting body, which can engulf them as it grows. The fruiting bodies are shelf-like and very irregular or totally flat on the substrate, and their size varies from just a few millimeters to 10–15 cm. In some cases, only small white mounds about 5 mm in diameter are found. These have a consistency of a pencil eraser, although firm pressure can crush them, and they represent a very early or an interrupted stage of development. Flat fruiting bodies form on the roofs of animal burrows within the infected sites and on roots exposed by the digging of trenches or soil pits.

Life history: Spores are liberated from the pores on the undersurfaces of the fruiting bodies and dispersed by the wind. Spore liberation occurs throughout the growing season but is greatest in the fall. Viable spores that land on a fresh stump of a host species germinate and initiate infection. Stump tops are very favorable sites for germination of *H. annosum* spores for about 4 weeks after cutting, and only a few other spores germinate in the first 2 weeks. Some spores are washed down through the soil and can initiate infection directly in the roots. Once infection is established, the fungus colonizes the roots of the stump and can invade the roots of other trees or stumps at points of root grafting or root contacts. Observations in Ontario indicate that tree death from fomes root rot becomes evident in a plantation about 5 years after thinning.

Control: Because 90% or more of the infections start in fresh stump tops, control efforts are aimed at protecting this site from colonization by *H. annosum*. An application of a light layer of granular borax to the stump tops immediately after the tree is felled is the current control measure recommended in Ontario. Also, because spore liberation by this fungus is highest in the fall, thinning is discouraged during that time of the year. Spore production during the winter months is very low or absent. Protecting pine stumps by spraying them with spores of *Phlebiopsis gigantea* (Fr.) Jül. (Syn.: *Peniophora gigantea* (Fr. : Fr.) Massee) has been employed successfully as a biological control procedure on pine in Great Britain.

Additional information: In much of the early literature, this fungus was called *Fomes annosus* (Fr. : Fr.) Cooke. *Heterobasidion annosum* has an imperfect state called *Spiniger meineckellus* (A. Olson) Stalpers. This state of the fungus is readily recognized with a hand lens and is formed by germination of spores of *H. annosum*.

Selected bibliography

Myren, D.T. 1973. The influence of experimental conditions on a test of borax and sodium nitrite as a stump protectant against infection by *Fomes annosus*. Environ. Can., Can. For. Serv., Sault Ste. Marie, Ont. Inf. Rep. O-X-191. 13 p.

Ross, E.W. 1973. *Fomes annosus* in the southeastern United States. U.S. Dep. Agric., For. Serv. Tech. Bull. No. 1459.

Prepared by D.T. Myren.

A

B

C

Plate 75

A. Red pine killed by *Heterobasidion annosum*, the causal agent of fomes root rot. Note pine with dead needles still attached and stumps of infected trees removed in an earlier sanitation operation.

B. Active fruiting bodies of *Heterobasidion annosum*, the causal agent of fomes root rot, on an infected red pine stump.

C. Small fruiting bodies of *Heterobasidion annosum* on red pine. The needle layer was removed to reveal these structures, often the only sign that fomes root rot is in the tree.

76. Tomentosus root rot
Inonotus tomentosus (Fr. : Fr.) Teng
Plate 76

Hosts: Mainly black, Norway, and white spruce; occasionally balsam fir and eastern white, jack, and red pine.

Distribution: Associated primarily with spruce in the Maritime provinces, Ontario, and Quebec.

Effects on hosts: This fungus causes a white pocket rot in the roots and lower stem of its host. The rot can extend up the stem as much as 2 m, causing significant volume loss in the butt log. Root decay results in reduced growth, mortality, and windthrow. Lowered vigor makes the host more

A

B

C

Plate 76

A. Fruiting bodies of *Inonotus tomentosus*, the red butt rot fungus, on a living white spruce.

B. Fresh fruiting body of *Inonotus tomentosus* on the forest floor in a white spruce plantation. The sporophore over the root system is often the only sign of red butt rot in the tree.

C. An older fruiting body of the red butt rot fungus, *Inonotus tomentosus*, showing the profile and somewhat faded color. (Photograph courtesy of R.D. Whitney.)

susceptible to damage from other pests and abiotic factors.

Identifying features: The fruiting bodies can be sessile on the trunk of the host but most often are stemmed and formed in the late summer or fall on the ground above infected roots. They are often numerous and may be found on all sides of the host. The fruiting structure is tan to yellow-brown and slightly velvet-like on the upper surface when young. On the lower surface, the fungus is poroid and light tan to brown, but it darkens where bruised. The caps of the fruiting bodies range in size from 3 to 18 cm in diameter and from 0.3 to 4 cm in thickness. The stems can be up to 5 cm in length and 0.5–2 cm in diameter. Pores extend down the stem for a short distance.

The decay is first evident as a red stain and later as small pockets that are lined with white fibers. Resin exudation is evident on infected roots.

Life history: Spores liberated from fruiting bodies are wind-borne and are responsible for infection. Infection appears to occur at or below ground level and is associated with a dead or wounded root or a stem wound. Once infection has occurred, the fungus moves out on the roots and can pass to other roots at points of grafting or contact. The stem of the host can be partially girdled below ground level, and the fungus moves into the tap root and other major laterals. This manner of spread results in the development of infection pockets, also referred to as infection centers. Infected trees may live for a number of years, often 15 or more, before death occurs. Fruiting occurs annually, often for a number of years on the same host.

Control: Trees in infected plantations should be harvested as soon as is practical. The site should be replanted with hardwoods or with conifers in which the disease does not appear to spread as quickly, such as balsam fir or pine.

Additional information: There are several names currently proposed for this fungus, and some researchers feel there are either two varieties or two separate species. The two-species concept, with *Inonotus circinatus* (Fr.) Gilbertson being the other species, seems to be more widely accepted. In early literature, the fungus was called *Polyporus tomentosus* Fr. : Fr., and *Polyporus circinatus* (Fr.) Fr. was considered similar or relegated to a variety.

Samples should consist of root sections with both decayed and apparently sound wood present. When possible, sampling should be done in the fall, when fruiting bodies are formed and can be included with the infected root sections.

Inonotus tomentosus is commonly referred to as the red butt rot fungus.

Selected bibliography

Whitney, R.D. 1962. Studies in forest pathology. XXIV. *Polyporus tomentosus* Fr. as a major factor in stand-opening disease of white spruce. Can. J. Bot. 40:1631-1658.

Whitney, R.D. 1977. *Polyporus tomentosus* root rot of conifers. Dep. Fish. Environ., Can. For. Serv., Great Lakes For. Cent., Sault Ste. Marie, Ont. For. Tech. Rep. No. 18. 12 p.

Prepared by D.T. Myren.

---

77. Brown cubical rot
Phaeolus schweinitzii (Fr. : Fr.) Pat.
Plate 77

Hosts: Mainly balsam fir and white and black spruce; occasionally eastern white and jack pine; rarely larch and hemlock.

Distribution: Common in mature and overmature stands of its hosts throughout eastern Canada.

Effects on hosts: Brown cubical rot is responsible for considerable volume loss in conifers, although it has been less important in recent years in eastern Canada. It is usually associated with mature and overmature stands, but young trees can be damaged and even killed by this fungus. It produces a rot in the heartwood of the roots and butt, often extending up as high as 3 m. As a result, infected trees suffer a significant loss in volume and are also more subject to windthrow.

Identifying features: The fruiting body of the fungus is annual. It may be shelf-like when it comes from wounds at the base of the tree or stemmed and somewhat funnel-like when produced on the ground around the tree. Fruiting bodies produced on the ground develop from infected roots and are the most common form of fruiting body in Ontario. From a top view, the stemmed form is roughly circular, slightly depressed in the center, reddish brown with a yellow margin, and velvety pubescent when fresh. On older fruiting bodies, the pubescence is reduced or lost as a result of weathering. The undersurface is poroid, yellowish green, and darkens when bruised. Internally, the fruiting body is yellow to reddish brown. The decay changes the color of the wood to dark reddish brown and causes it to break into cubical blocks. Thin mats of white fungal growth are sometimes seen between the blocks.

Life history: Fruiting bodies develop from infected hosts in the late summer and fall, being more numerous during wet periods. Spores are produced on the walls of the pores on the lower surface and are dispersed by winds when liberated. Infection occurs in basal wounds, particularly fire scars.

Control: Control of decay fungi is often difficult, but the prevention of fire and early harvesting of mature trees should help mitigate the damage caused by this fungus.

Additional information: Studies in the boreal forests of Ontario have shown *Coniophora puteana* (Schumacher : Fr.) P. Karsten and *Serpula himantioides* (Fr. : Fr.) P. Karsten to be the two most commonly encountered brown rots. *Phaeolus schweinitzii* was almost totally absent from these studies. In early literature, *P. schweinitzii* was placed in the genus *Polyporus*.

Selected bibliography

Boyce, J.S. 1961. Forest pathology. 3rd ed. McGraw-Hill Book Co., New York, NY. 572 p.

Hepting, G.H. 1971. Diseases of forest and shade trees of the United States. U.S. Dep. Agric., For. Serv. Agric. Handb. No. 386. 658 p.

Prepared by D.T. Myren.

Plate 77

A. The undersurface of an active fruiting body of *Phaeolus schweinitzii*, the cause of brown cubical rot, collected from a balsam fir root.

B. The upper surface of an active fruiting body of *Phaeolus schweinitzii*, the cause of brown cubical rot, collected from a balsam fir root.

A

B

78. **Rhizina root rot**
***Rhizina undulata* Fr. : Fr.**
Plate 78

Hosts: Species of spruce and pine.

Distribution: Only in burned areas in New Brunswick, Newfoundland, Ontario, and Quebec.

Effects on hosts: The fungus causes decay of roots and results in the mortality of groups of seedlings up to 5 years of age in regeneration or young plantations established on or around old fire sites. The damage is more prevalent in the first year after a burn, reduced in the second year, and disappears in the third year.

Identifying features: The most conspicuous above-ground symptoms of the disease are discoloration of needles, sparse foliage, thin crown, and resinous exudations from the lower trunk — similar to those of other root rots or drought. Close examination of the roots of dead seedlings shows infected roots closely matted together with a mass of white or yellowish mycelial strands with characteristic angles, clinging to or penetrating through lenticels, and ramifying through the cortical tissue of the root. The attacked lenticels often become resin-filled and appear as white spots on the roots. The mycelium can often be traced through soil to nearby fruiting bodies of rhizina. The fruiting bodies indicate the presence of the fungus. They are produced on the mineral soil or forest duff of burned trees in sheltered areas in coniferous forests with rhizoids attached to wounded or severed roots and buried woody debris. They appear first as nickel-sized brown buttons with a yellow edge. The mature, discoid, convex to convoluted, brain-like, irregularly shaped fruiting bodies are 5–12 cm in diameter and have undulating, pale brown to yellow-brown upper surface with narrow white margin. Old fruiting bodies are almost black. They often occur in crust-like groups or lines following roots. The fruiting bodies are supported by root-like growths — the rhizoids — that connect the fruiting body with woody material or roots of living or dead trees.

In clear-cut areas that are burned, fruiting bodies must be produced 10–16 months after burning if they are to develop.

Plate 78

Fruiting bodies of *Rhizina undulata*, the cause of rhizina root rot, on a site burned early in the preceding year. (Photograph courtesy of H.L. Gross.)

Life history: The fungus appears to be generally present in soils of coniferous forests. The fruiting bodies are annual, usually appearing in early summer and often persisting until frost in the fall. Spores produced in these fruiting bodies are dispersed by wind in summer and are washed into soil by rain. They can survive in soil for at least 2 years under field conditions. Fires destroy many spores, but some survive, germinate, and colonize duff and living roots of conifers. These roots constitute a food base for the fungus, enable the fungus to spread out radially by means of cream or yellow mycelial strands, and infect and kill further seedlings. After successful invasion, adjacent roots are killed; as the fungus progresses, the cortical tissue is destroyed, and the seedling is killed. Later, the fruiting bodies are formed at or close to the base of the infected seedling, and the life cycle of the fungus is repeated.

Once the mycelium of the fungus is established in the roots, mortality of seedlings begins, and the fungus spreads to nearby seedlings, often without the formation of more fruiting bodies or the occurrence of fire.

Control: Because the economic importance of the disease has not been assessed, there is no recommended direct control of rhizina root rot. However, the problem can be avoided or reduced by banning bonfires in forests, particularly on sites that may be used for planting in future, and delaying reforestation of burned infected sites by at least 2 years.

Additional information: This fungus has been referred to as the tea pot fungus because it was found to start at points where fire was used by forest workers to boil water for tea. *Rhizina inflata* (Schaeffer) P. Karsten is another name used for this fungus.

Selected bibliography

Baranyay, J.A. 1972. *Rhizina* root rot of conifers. Environ. Can., Can. For. Serv., Pac. For. Res. Cent., Victoria, B.C. For. Insect Dis. Surv. Pest Leaflet No. 56. 5 p.

Boyce, J.S. 1961. Forest pathology. 3rd ed. McGraw-Hill Book Co., New York, NY. 572 p.

Ginns, J.H. 1968. *Rhizina undulata* pathogenic on Douglas-fir seedlings in western North America. Plant Dis. Rep. 52:579-580.

Peace, T.R. 1962. Pathology of trees and shrubs with special reference to Britain. Oxford University Press, London. 753 p.

Prepared by Pritam Singh.

79. Spruce cone rust
Chrysomyxa pirolata (Körn.) Winter
Plate 79

Hosts: Mainly black and white spruce; occasionally blue spruce; alternate hosts are pyrolas and single-delight.

Distribution: Common throughout eastern Canada.

Effects on hosts: The disease causes malformation, premature opening and destruction of cones, reduced seed production, and decreased viability of seeds. Seeds usually fail to develop in affected cones, and those that do may germinate abnormally. In general, a cone is either completely infected or completely sound. In epidemic periods, the fungus may cause severe damage over fairly large areas.

Identifying features: The most conspicuous symptoms of the disease are small, yellow-orange spots

A

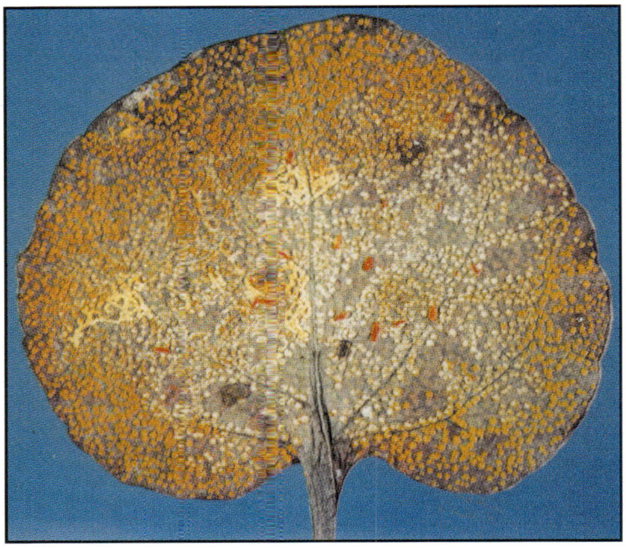

C

B

Plate 79

A. *Chrysomyxa pirolata*, the cause of spruce cone rust, fruiting on black spruce.

B. *Chrysomyxa pirolata*, the cause of spruce cone rust, fruiting on Norway spruce.

C. *Chrysomyxa pirolata*, the cause of spruce cone rust, fruiting on pyrola, the alternate host for the rust fungus.

on young cones. In late summer, the rusted, prematurely brown, and opened cones show conspicuous yellow or orange-yellow powdery masses of rust spores on the surface and around the edges of the cone scales.

The disease causes slight atrophy and yellowing of the foliage of the alternate hosts.

Life history: The fungus requires two different host species to complete its life cycle: a conifer and a herbaceous weed (pyrolas or single-delight). Spores from cones do not spread the disease to other cones but only to the alternate hosts. In late spring or early summer, two kinds of yellow-orange, spore-producing pustules appear on the undersides of the leaves of the alternate host. Spores from one kind of pustule persist throughout the growing season and spread the disease to other pyrolas or single-delight; a second type of pustule produces spores that germinate and produce a finer and smaller type of spore that infects the spruce cones. Cones appear to be susceptible to infection shortly before, during, and after pollination.

Control: The disease can be controlled by eradication of alternate hosts around the seed orchard, by application of fungicides to cones, or both.

Additional information: Infection of the cones can be both systemic and partly systemic in the herbaceous host. Infected cones have been found to be more attractive than uninfected cones to some cone-attacking insects.

Selected bibliography

Singh, P. 1981. Inland spruce cone rust of black spruce in Newfoundland and Labrador. Can. Plant Dis. Surv. 61(2):43-47.
Ziller, W.G. 1974. The tree rusts of western Canada. Environ. Can., Can. For. Serv., Victoria, B.C. Publ. No. 1329. 272 p.

Prepared by Pritam Singh.

80. Animals: 1
Mice, rabbits, porcupines, and beavers
Plate 80

Hosts: Miscellaneous.

Distribution: Widespread in eastern Canada.

Effects on hosts: The bark of trees is removed by these animals, resulting in wounds; if girdled, the tree dies. Beavers often fell trees.

Identifying features: Damage done in winter is easily detected during midsummer when an entire conifer turns red or when the new foliage of hardwood trees dies soon after flushing or does not develop at all. To complete the diagnosis, trunks of trees must be checked for wounds on the bark. The animal responsible may be identified by the size, location, and portion of the tree attacked. In some cases, actual tooth marks can be seen, and the animal can be identified from these. Usually, rabbits and mice prefer small trees, whereas porcupines and beavers damage larger trees.

Biology of damage: Mice and rabbits feed on bark during the winter. Mice travel beneath the snow, and rabbits often seek shelter among the trees upon which they are feeding. Porcupines climb in trees and eat patches of bark at any height. Beavers cut down entire trees to get at the tender bark and branches at the top and to obtain material with which to construct dams and shelters. Porcupines and beavers are usually active only during the growing season and late fall, although porcupines also feed during milder parts of the winter.

Control: Mouse control in plantations starts with a weed control program that reduces suitable habitat. Hardware screen can be placed around each tree to prevent damage by both mice and rabbits. Hunting rabbits is also an effective control measure. Animal repellents can be used but are not always effective. Poison bait is employed in some cases, but pesticide regulations should be checked before this method of control is used. Porcupines are quite difficult to control, even by hunting. Beaver populations can be well controlled by trappers.

Additional information: Animals such as deer, moose, cattle, and horses (see Section 81) may also damage trees. Domestic animals are usually controlled by keeping them within enclosures. Wild animals are much more difficult to control. A few squirrel species strip the outer bark from hardwoods and feed on the inner bark (see Section 81). A food shortage or the quality of sap seems to cause this type of feeding. Squirrels can also cause damage to trees by tearing off cones.

Selected bibliography
Tattar, T.A. 1978. Diseases of shade trees. Academic Press, New York, NY. 361 p.

von Althen, F.W. 1983. Animal damage to hardwood regeneration and its prevention in plantations and woodlots of southern Ontario. Environ. Can., Can. For. Serv., Great Lakes For. Cent., Sault Ste. Marie, Ont. Inf. Rep. 0-X-351. 28 p.

Prepared by G. Laflamme.

A

B

Plate 80

A. Aspen felled by beaver activity. Gnawing on the stems is very characteristic and can be readily identified.

B. Extensive mortality in a Scots pine plantation caused by mice feeding on the bark and girdling the stem.

C. Young sugar maple girdled by mice feeding on the bark.

D. Young tamarack girdled by rabbits feeding on the bark.

E. Jack pine killed by porcupines feeding on the bark.

F. Damage to jack pine caused by porcupine feeding.

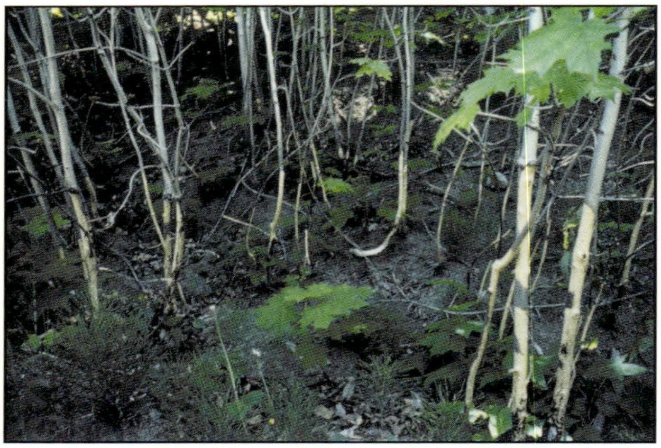

C

D

E

F

81. Animals: 2
Squirrels, cattle, and birds
Plate 81

Hosts: Miscellaneous.

Distribution: Widespread.

Effects on hosts: Damage done by squirrels to twigs and by pine grosbeaks to buds does not endanger the life of trees but can cause twig death and the formation of multiple leaders. Rows of peck holes made by sapsuckers gradually kill the bark and may girdle a tree. Soil compaction and wounds inflicted to trunk and roots by cattle and horses promote tree decline, which eventually leads to the death of the tree.

Identifying features: Squirrel damage in pine stands and plantations is detected by "red flags," which are twigs killed by squirrels when they tear away pine cones that were firmly attached to the tree. Squirrels also damage balsam fir by clipping those twigs that bear male flowers, eating the flowers, and letting the twigs fall to the ground. Fallen twigs are often numerous and easily detected on the snow at the base of a tree. The pine grosbeak eats the succulent part of pine buds, leaving the scales. These birds usually feed in pine plantations, and the damage is localized. Peckholes made by sapsuckers can be recognized as regular rows of holes in the bark. The bird prefers birch, maple, hemlock, and Austrian pine but may feed on other tree species. Patches of bark can be killed; if girdled, the damaged tree can die. Paths used by cattle and horses in plantations or forests are easily detectable by soil compaction and the lack of regeneration. Animal grazing not only destroys the existing trees but affects regeneration as well. Grazed areas cease to be productive and develop into relatively barren forest land, sometimes populated by undesirable plant species.

Biology of damage: Squirrels and pine grosbeaks damage trees in the fall and winter, usually when there is a shortage of food. Sapsuckers are active during the growing season, coming back repeatedly to feed on the sap accumulated in the holes they made and making new holes if the host is to their liking. Woodpeckers may cause some damage, but their impact is not severe because their drill holes are made to get at insects in the wood of dead or dying trees. They occasionally go after insects in trees that would withstand the insect damage, however, and in these cases they may be a problem.

Control: Because squirrels usually do not damage trees in the same area 2 years in a row, no control is necessary. Pine grosbeaks are an annual problem in some pine plantations and are hard to control. Some success has been achieved with automatic noisemakers. Sapsuckers are also difficult to control. However, their potential nesting trees, such as large decayed poplar, could be removed. Birds may be frightened by aluminum plates, disks, or bands hanging on branches of their favorite trees. Rubber snakes hanging in trees have been reported to be successful. Cattle and horses should be kept in enclosures outside forested areas and tree plantations.

Additional information: There are other animals that can cause damage to trees. Rabbits, deer, and moose eat tender twigs. A high population of deer may reduce or change the species composition of seedlings or even cause a complete regeneration failure. They seem to be particularly fond of jack pine. Bears often claw on bark, causing severe wounds. Male dogs marking cedars often kill the lower branches of ornamentals. The foliage killed by animal urine usually has a blackish cast.

Selected reference
Tattar, T.A. 1978. Diseases of shade trees. Academic Press, New York, NY. 361 p.

Prepared by G. Laflamme.

A

B

C

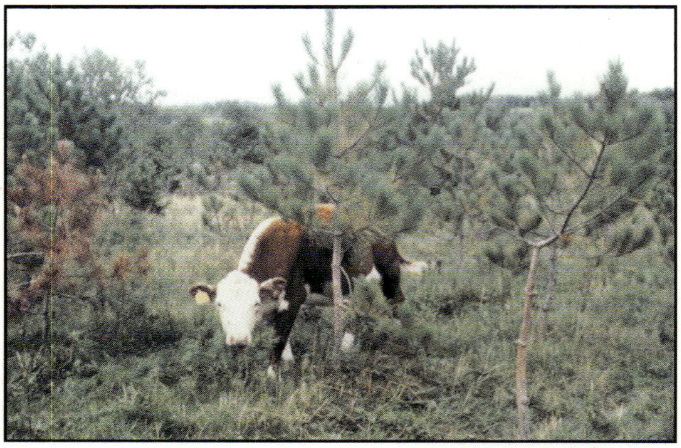

D

E

F

G

Plate 81

A. Squirrel damage to jack pine. Note the old scar at the base of the dead shoot where a cone was torn off.

B. Recent injury to jack pine by a squirrel tearing off a cone.

C. Damage to red pine by deer browsing.

D. A young pine plantation damaged by livestock browsing, branch breakage, and soil compaction.

E. Sapsucker damage on birch. The pattern of the feeding holes is very diagnostic. Birch is a favorite host from midsummer to early fall.

F. Sapsucker damage to balsam fir. Although balsam fir is not a favorite host, considerable damage can be caused to trees selected for feeding.

G. Woodpecker damage to sugar maple.

82. Leaf spots and galls
Mites and insects
Plate 82

Hosts: Hardwoods.

Distribution: Widespread.

Effects on hosts: Heavy infestations may affect the appearance of a tree but do not endanger its life.

Identifying features: Attack by mites and insects often causes a variety of symptoms, but the most common ones are discolored spots and galls. These are commonly observed on oak and maple leaves. On maple, the most commonly reported symptoms are a yellow spot surrounded by a red circle, a purple velvet spot, and a spindle-shaped gall. On oak, galls are much more common.

Life history: Because of the large number of different mites and insects responsible for such damage, it is difficult to describe their individual life cycles. Our main purpose here is to make one aware of them and the symptoms of their damage, which can be very similar to those caused by fungi.

Control: Measures are not usually required to control spots and galls caused by insects and mites, as primarily the aesthetics of a tree are affected. In the case of heavy infestation, a suitable insecticide could be used to reduce the population of the pests, but this would not reduce the damages of the current year. Protection usually requires the pesticide to be applied before damage is evident, and the decision to do so must be based on the level of damage the previous year. Also, as insects and mites are at the mercy of the environment, there is no way to predict with certainty if the pest population will be significant as the year progresses.

Additional information: There are numerous other "leaf spot-like" discolorations that may be observed, some of which would be caused by nonpathogenic conditions.

Selected bibliography
Rose, A.H.; Lindquist, O.H. 1982. Insects of eastern hardwood trees. Environ. Can., Ottawa, Ont. For. Tech. Rep. 29. 304 p.
Tattar, T.A. 1978. Diseases of shade trees. Academic Press, New York, NY. 361 p.

Prepared by G. Laflamme.

A

B

C

D

E

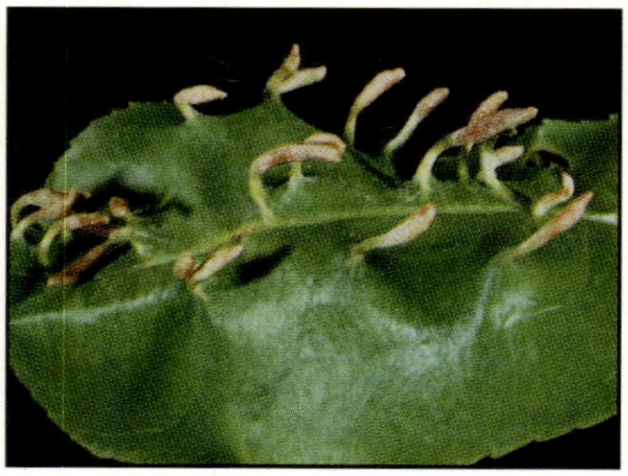

F

Plate 82

A. Leaf spot damage on balsam poplar caused by a midge (*Contarinia* sp.).This insect-caused spot is often attributed to a fungus.

B. Leaf spot on red maple caused by *Acericecis ocellaris*. This insect-caused leaf spot is often confused with the fungus *Phyllosticta minima*.

C. Malformation on a basswood leaf caused by the linden wartgall midge *Cecidomyia verrucicola*.

D. This red coating on sugar maple leaf is caused by the mite *Aceria regulus*. It is common but considered relatively harmless.

E. Typical discoloration of maple by the mite *Aceria elongatus*. It is not considered to be damaging.

F. Malformation of a cherry leaf caused by the mite *Eriophyes padi*. Similar malformations are seen on maple and other hosts but cause no significant injury.

83. **Air pollution**
Plate 83

Tree damage from air pollution is a recognized problem in eastern Canada but at this time seems confined primarily to areas around those industries that are the sources of pollutant emissions. Certainly, long-range transport of air pollutants does occur, but the extent of their impact on the forests of eastern Canada has not yet been established. Considerable effort is being expended on research to answer this question.

Pollution damage to trees is detected by observation of the symptoms on the foliage. Foliage damage can result in reduced growth rate and reduced vigor. If severe, air pollution can result in plant death. The symptoms expressed by the foliage vary between hosts, between pollutants, with pollutant concentration, with distance from the source, and with duration of exposure. Normally the damage extends much farther because of the prevailing wind, although certain areas may have less damage or may even escape damage because of wind patterns or their topographical positions. The major pollutants and the symptoms they cause are discussed very briefly below.

One of the most common air pollutants is sulfur dioxide (SO_2). It is produced by the combustion of fossil fuels, primarily coal. Much of the coal consumed is used by power plants producing electricity. Smelters are also common sources of SO_2 emissions. A number of other industrial processes also use fossil fuels and contribute to the total SO_2 content of the atmosphere. The damage caused by SO_2 usually involves yellowing of older needles on conifers and discoloration, interveinal necrosis, and defoliation on hardwoods. Conifer trees in general are more severely injured. Frequently the foliage of affected trees is smaller than that of healthy trees and is shed earlier. Seedlings can be killed and cone crops reduced, which would interfere with regeneration in polluted areas.

Fluoride is also an important air pollutant and is primarily released into the atmosphere from metal mining and smelting, brick manufacture, phosphate fertilizer production, and several other industrial processes. Fluoride causes a tip burn on the current year's needles on pine and eventually needle death. Older needles seem to be quite resistant to fluoride damage. Hardwood leaves exhibit tip and marginal necrosis, which eventually involves the entire leaf.

Ozone (O_3) is also an important cause of plant damage and is a natural component of the atmosphere. Most of the O_3 causing damage to plants comes from automobile emissions. Ozone causes green and yellow mottling on conifer needles and reddish to brown flecks on hardwood leaves. The color of the flecking varies between plant species.

Another product of automobile emissions is called peroxyacetyl nitrate (PAN). PAN causes a bronzing or silvering on the undersides of hardwood leaves. PAN damage to conifers in eastern Canada has not been well described or documented.

Nitrogen oxides (NO_x) are produced by automobiles, power generating stations, and a number of industrial processes. Conifers damaged by NO_x show dead needle tips or entire needles. Hardwood leaves usually show a marginal yellowing and necrosis between veins. The above discussions dealt with five of the more important pollutants that damage trees. Ammonia, chloride, and ethylene also play roles as air pollutants. These chemicals are produced by a variety of manufacturing processes. They cause tip or total needle death on conifers and discoloration of hardwood leaves.

Acid rain, more correctly called acid deposition, is the result of the long-range transport of airborne pollutants (LRTAP). The pollutants are sulfur dioxide (SO_2) and oxides of nitrogen (NO_x), which are carried long distances and are transformed to sulfates and nitrates. These chemicals combine with moisture in the atmosphere and are carried down by rain or snow as weak solutions of sulfuric acid and nitric acid. Although the specific impact of acid rain on the forest is under investigation, it is strongly suspected that it is responsible for discoloration of foliage, defoliation, crown dieback, and possibly eventual death of trees.

Control requires establishment of plantations in areas beyond that influenced by a particular pollution source and use of trees that are more pollution-tolerant where conditions warrant. Removal of dead shoots and trees is recommended to prevent buildup of secondary pests.

Selected bibliography

Addison, P.A.; Rennie, P.J. 1988. The Canadian Forestry Service air pollution program and bibliography. For. Can., For. Sci. Dir., Ottawa, Ont. Inf. Rep. DPC-X-26. 133 p.

Anonymous. 1982. Downwind, the acid rain story. Environ. Can., Ottawa, Ont.

Loomis, R.C.; Padgett, W.H. 1974. Air pollution and trees in the east. U.S. Dep. Agric., For. Serv., State Priv. For., Atlanta, GA. 28 p.

Malhotra, S.S.; Blavel, R.A. 1980. Diagnosis of air pollutant and natural stress symptoms on forest vegetation in western Canada. Environ. Can., Can. For. Serv., Edmonton, Alta. Inf. Rep. NOR-X-228. 84 p.

Prepared by D.T. Myren and Pritam Singh.

A

B

C

D

E

Plate 83

A. Asbestos dust on a balsam fir near a plant where asbestos products were manufactured.

B. Chlorine damage on spruce seedlings resulting from excessive use of commercial bleach to clean adjacent equipment.

C. Ozone damage to an eastern white pine. Note necrosis of needle tips.

D. Sulfur dioxide damage to a forested area near a coal-burning power generating facility.

E. Sulfur dioxide damage to spruce seedlings.

84. **Herbicides**
Plate 84

Herbicides are chemicals used for killing unwanted plants. Some herbicides are very selective, killing only certain kinds of plants, whereas others kill all or most vegetation in a treated area. Many homeowners employ herbicides for weed control in their lawns, and herbicides are widely used in forestry and agriculture. Herbicides are commonly used in forestry to keep fire lanes and bush roads open, to clear powerline corridors, and to control weeds in young pine plantations, for example.

Damage from improper or indiscriminate use of herbicides is very common in eastern Canada. Improper dosage, improper timing, and improper application often cause damage, but diagnosis can be difficult. Perhaps the most easily recognized damage is that caused by herbicides selective for broadleaf plants, such as dandelion, drifting to broadleaf ornamentals. The leaves of ornamental trees will cup and twist and often develop whip-like leaf tips. Defoliation may result, but many trees

A

B

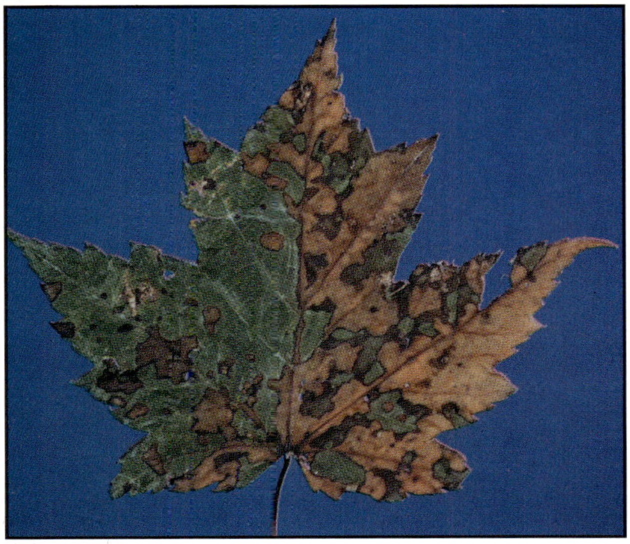

C

Plate 84

A. Herbicide damage to ash. The herbicide was being used on a lawn in which the tree was planted.

B. Herbicide damage to spruce. Damage resulted from drift of a herbicide incorporating 2,4-D.

C. Maple damaged by herbicide drift.

do recover. Herbicides can also be picked up and translocated by the root system, causing damage over much of the affected plants. Damage to part of the crown is often noted when only one or two of the major roots take up the poison.

Herbicide damage is common in Christmas tree plantations when the chemical is sprayed directly on newly developing needles and also at the edge of agricultural fields where herbicides are used. In the latter situation, damage may result from drift or root translocation and has been noted when potatoes or corn is the adjacent crop.

Prevention of herbicide damage requires following the manufacturers' directions carefully and using the herbicide only when necessary and only for the purposes stated on the container. Spraying should be avoided on windy days when drift is likely or when the temperature is not within the limits recommended by the manufacturers. Equipment used to apply herbicides should not be used for application of any other chemicals to broadleaf plants. When ornamental trees are present, application of a herbicide under the drip line should be kept to a minimum. This is particularly important if the herbicide is being applied in a mix with fertilizer. Christmas tree plantations should not be established near areas where herbicides are used extensively, such as power lines and roadsides. Trees killed or damaged by herbicides should be removed to reduce the buildup of fungi and insects that may attack neighboring healthy trees.

Selected bibliography

Alex, J.F.; Waywell, C.G.; Switzer, C.M. No date given. Weed control in lawns and gardens. Ont. Minist. Agric. Food, Toronto, Ont. Publ. No. 529. 100 p.

Cordukes, W.E. 1979. Home lawns. Agric. Can., Ottawa, Ont. Publ. No. 1685E. 27 p.

Prepared by D.T. Myren and Pritam Singh.

85. **Insecticides**
Plate 85

Two commonly used insecticides have been observed to cause damage to ornamental trees. Malathion, used to control spruce budworm, was observed to cause black spots on adjacent maples. This insecticide can be used on maples, and it is suspected that the spotting was a result of an inappropriate concentration of the chemical, an inappropriate application time, or both. Dimethoate is commonly used on birch to control birch leaf miner. Excessive use of this insecticide as a foliage spray has caused browning and yellowing of the tissue between major veins of the leaves and on the leaf margins.

We do not recommend that these insecticides not be used, but we do strongly recommend that manufacturers' directions be followed carefully, and we urge all users of pesticides to be very careful in their preparations for use.

Prepared by D.T. Myren and L.P. Magasi.

Plate 85

A. Damage to birch from excessive use of the insecticide dimethoate, shortly after application.

B. Damage to birch from excessive use of the insecticide dimethoate earlier in the summer.

A

B

86. **Ocean spray**
Plate 86

Damage from saltwater spray is found on the coastal areas of eastern Canada and on the island of Newfoundland. Both conifers and hardwoods suffer from this form of salt injury. Salt spray is carried by strong winds, and damage on the exposed side of trees can be seen many kilometers inland. The most conspicuous effect is the death of foliage from salt deposition. Conifers appear as though damaged by fire, becoming bright orange-red. This symptom is particularly noticeable on eastern white pine and balsam fir. On hardwoods, the marginal and interveinal areas appear scorched on the exposed sides of the trees. Repeated or continuous exposure to ocean spray can cause mortality.

There is no direct control for this type of damage. Salt-tolerant species should be used near the ocean and could be used as shelterbelts to protect plantations of less tolerant species.

Selected reference

Peace, T.R. 1962. Pathology of trees and shrubs with special reference to Britain. Oxford University Press, London. 753 p.

Prepared by Pritam Singh.

Plate 86

Salt damage to pines from ocean spray.

87. **Salt**
Plate 87

Damage to roadside trees and shrubs by de-icing salt is common in areas of significant snowfall. The salt causes damage to conifers by splash that lands on needles, and to both conifers and hardwoods through buildup in the soil at points where salt-laden water tends to accumulate. Conifers usually show most damage on the side of the tree facing the road and on the distal half of the needles when splash is the cause. Buds are not usually damaged; the new foliage and loss of injured foliage often give the conifers a somewhat normal appearance by midsummer. Damage from salt accumulation in the soil can kill some conifers and can cause significant twig dieback on hardwoods. Leaves of hardwoods that have taken up salt through the roots often show browning on the margins.

Recognition of salt damage can be difficult if samples are collected well into the growing season, but it should be kept in mind whenever damage is noted along highways and streets in populated areas. The use of salt-tolerant trees, planting farther back from the road, and avoiding areas where ponding occurs help reduce the problem.

Selected reference

Manion, P.D. 1981. Tree disease concepts. Prentice-Hall, Englewood Cliffs, NJ. 399 p.

Prepared by Pritam Singh and D.T. Myren.

Plate 87

Salt damage to pines from road de-icing.

88. **Frost**
Plate 88

Frost is a commonly encountered problem in eastern Canada and can have serious effects on tree growth. There are three types of frost injury: shoot mortality, frost heaving, and frost cracks.

Shoot mortality: Young balsam fir and spruce are particularly susceptible to shoot damage from late frosts. Other conifers and hardwoods can also be injured by frost. Damage is seen when a late frost occurs after bud break, when new succulent growth is present. The young tissue is killed; if this occurs for several successive years, stunted or bushy trees result. Smaller seedlings can be killed when frost is severe. Symptoms of frost damage include curling, death, and reddening of all or part of the new shoots. The dead buds often remain on the affected shoots until late fall, and some persist until the following spring. Dead, overwintered buds are usually quite shriveled and blackish in color.

Frost damage to hardwoods is usually confined to tender new leaves, which turn black, shrivel and pucker, and drop prematurely. Older leaves often show marginal browning but usually stay on the tree throughout the growing season.

Selection of frost-hardy planting stock and of sites less prone to frost helps reduce damage. Frost is often most severe in low-lying areas and occurs repeatedly on these sites. Such frost-prone areas are often referred to as frost pockets.

A

B

Plate 88

A. Late frost damage to spruce. Balsam fir also suffers from late frosts with similar damage.

B. Frost crack on a hardwood. Conifers also suffer from frost cracks.

Frost heaving: Heaving is the uprooting of seedlings caused by repeated freezing and thawing of the soil. The seedling is lifted so that the root collar is above the soil surface or may even be thrown completely out of the ground. The roots often break off. Frost heaving is most common on heavy soils with high proportions of clay. Provision of brush, ground cover, or mulching helps to reduce frost heaving. In planting areas where this is a potential problem, it is best to leave the organic matter on the ground to provide some protection against the rapid and frequent temperature changes.

Frost cracks: Frost cracks are radial splits of tree trunks. They occur during winter when there is a sudden and pronounced drop in temperature. The splits originate at the base of the trunk and extend upwards for several meters. The cracks often close in warmer weather, and the tree responds to the wound by producing callus. Sometimes healed cracks open again, and a ridge called a frost rib develops after several seasons of healing and cracking. Frost cracks do not cause serious damage, but they do open the tree to invasion by stain and decay fungi. The physical splitting is responsible for "gun shots" heard in the forest on cold winter days. Both hardwoods and softwoods are susceptible to this form of injury.

Selected bibliography

Boyce, J.S. 1961. Forest pathology. 3rd ed. McGraw-Hill Book Co., New York, NY. 572 p.

Peace, T.R. 1962. Pathology of trees and shrubs with special reference to Britain. Oxford University Press, London. 753 p.

Prepared by Pritam Singh.

89. **Hail**
Plate 89

Hailstorms are not uncommon in eastern Canada and occasionally are so severe that significant damage to trees occurs. The damage caused may be superficial but can result in deep laceration of bark and stem tissue with defoliation or tearing of foliage and breakage of branches. Thin bark in the upper crown or on young shoots, developing buds, and young developing foliage are particularly subject to serious damage. Hail injury is relatively easy to diagnose, as the wounds are almost always on the upper sides of the branches on the side of the trees facing the storm. Trunk wounds face the same direction, and the same type of damage is found on adjacent trees. Scars from the injuries are evident for many years. Pitch accumulations beneath healed stem wounds can be identified in cross sections years later and allow the date of the hailstorm to be determined.

There is no control for hail damage. Ornamental trees could be pruned to eliminate broken branches resulting from hailstorms, and severely damaged trees should be removed.

Selected bibliography

Linzon, S.N. 1962. Hail damage to white pine and other trees. For. Chron. 38:497-504.

Riley, C.G. 1953. Hail damage in forest stands. For. Chron. 29:139-143.

Prepared by Pritam Singh and D.T. Myren.

Plate 89

Eastern white pine branches with scars from hail injury the previous year.

90. **High temperatures**
Plate 90

All plants are subject to damage from high temperatures. Four of the most common temperature problems encountered in eastern Canada are leaf scorch, late-spring leaf scorch, physiological needle droop, and winter sunscald.

Leaf scorch: Leaf scorch is common on hardwoods in eastern Canada and is caused by high temperatures accompanied by drying winds. Damage is most serious on exposed trees and on the side facing prevailing winds. Small trees can be killed, and ornamental trees can be rendered unsightly. Scorch occurs when leaves lose water faster than it can be replaced. Leaf margins and tissue between major veins die, turning light brown; when scorch is severe, leaves wilt and fall prematurely. Anthracnose (see Section 1) causes similar damage, and a laboratory examination is needed to differentiate this from scorch.

Scorch can occur on conifers but is not common. It is seen occasionally on nursery stock.

There is no control for leaf scorch. Watering during dry periods may help, but scorch can arise even when there is ample water in the soil. Planting trees in sheltered locations when possible would help reduce damage.

Late-spring leaf scorch: Late-spring leaf scorch (LSLS) occurs periodically in southern Ontario during a short period in early June. Damage occurs to immature leaves exposed to several days of overcast, wet weather followed quickly by bright sunshine and strong winds. Leaf margins turn brown, and entire leaves may be killed. Maple seems most susceptible to LSLS, but other trees can be damaged. This problem was noted in Sault Ste. Marie, Ontario, during mid-June in 1984. Many maples were damaged, and a few were totally defoliated.

There is no control for LSLS, but severely affected trees should be fertilized and kept well watered.

Physiological needle droop: Physiological needle droop is a condition caused by water loss from immature conifer needles at a rate exceeding replacement. This results in death of the succulent tissue under the needle sheath. This tissue collapses, and the needle hooks sharply downward. This condition occurs on red pine in Ontario every few years and is related to high temperatures and winds.

Winter sunscald: Winter sunscald results in the death of bark on the southwest face of trees. This occurs in the late winter or early spring as a result of daytime temperatures that raise bark temperature above freezing followed by a very rapid drop to temperatures below freezing at night. This form of injury has been noted on ornamental maples in Ontario and Quebec. Ornamental trees can be protected by the application of whitewash or tree wrap to the trunk in the fall. Paint should not be used for this purpose, as it may prove toxic to the tree.

Selected bibliography
Boyce, J.S. 1961. Forest pathology. 3rd ed. McGraw-Hill Book Co., New York, NY. 572 p.

Linzon, S.N.; McIlveen, W.D.; Pearson, R.G. 1972. Late-spring leaf scorch of maple and beech trees. Plant Dis. Rep. 56:526-530.

Prepared by Pritam Singh and D.T. Myren.

A

B

C

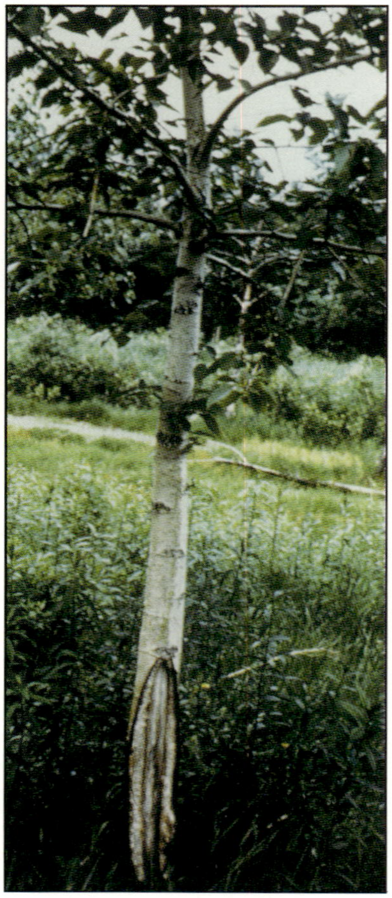

D

Plate 90

A. Leaf scorch on sugar maple.

B. Physiological needle droop of the current year on red pine.

C. Physiological needle droop on red pine the year following (B), showing the needle necrosis.

D. Sunscald canker on the south side of a hybrid poplar.

91. Lightning
Plate 91

Injury to trees by lightning is a fairly common occurrence in eastern Canada, although the total number of trees damaged is considered to be relatively low. All tree species can be damaged, with injury ranging from almost symptomless to complete shattering. A long strip of bark torn from the stem, often with a groove carved in the wood, and actual splitting are the most commonly observed forms of injury. Lightning is also an important cause of forest fires.

Damage to hardwoods tends to be limited to individual trees, whereas damage to conifers may occur to single trees or small groups of trees. The latter may result from multiple strikes or movement of electricity through root grafts. Branches on the pocket side of adjacent living trees may also show some damage. Trees in such pockets are quickly attacked by bark beetles and other insects. Unless damage by lightning is very evident, and it frequently is not, the cause of the mortality might be attributed to these secondary agents.

The extent of the damage determines the treatment to be used on damaged trees. Killed or badly damaged trees should be removed. This reduces the opportunity for a buildup of insect populations and prevents damage from further breakage by other physical agents, such as winds or snow. Pruning may be all that is required on trees with a small amount of damage, but the tree may be more severely damaged than it appears and may require additional pruning or possibly even removal in the future.

Trees can be equipped with lightning rods, but these are rarely worth the expense. Trees with lightning rods would be more likely to be struck but would suffer little damage if the rods were installed correctly. Installation of lightning rods should be done only by a professional arborist.

Selected bibliography

Harris, R.W. 1983. Arboriculture, care of trees, shrubs, and vines in the landscape. Prentice-Hall, Englewood Cliffs, NJ. 688 p.

Kourtz, P. 1967. Lightning behaviour and lightning fires in Canadian forests. Dep. For. Rural Dev., For. Branch, Ottawa, Ont. Publ. No. 1179. 33 p.

Prepared by D.T. Myren.

A

B

Plate 91

A. Extensive bark loss on a hardwood tree struck by lightning.

B. A small patch of spruce killed by lightning.

92. **Snow and ice**
Plate 92

Damage from snow and ice is common throughout eastern Canada, particularly in those areas that are subject to frequent and heavy snow and ice accumulations and winter storms. Many coniferous and broad-leaved trees are affected, but tall and wide-crowned trees are most susceptible to injury.

Heavy accumulations of snow or ice accompanied by high winds may uproot trees or break their stems below the crown. In less severe situations, only tree tops or leaders and branches break; the trees usually survive. The broken leaders are replaced by undamaged branches but result in deformed stems. In some cases, the tops and the branches do not break but become permanently bent and deformed. The damaged trees or shrubs are unsightly. Destruction of young cones and flower buds and stripping of bark are other forms of damage caused by snow and ice.

Not only do these injuries affect the normal growth and form of the tree, but the wounds created serve as entry points for decay fungi. Economically, damage can reduce yield and merchantability of affected trees and may necessitate replacement of ornamentals.

Very cold snow usually falls in small particles that do not readily cohere, whereas snow nearer its melting point falls in larger flakes that can build up on twigs and branches to such an extent that bending or breaking occurs.

Ice formation takes place when there is a substantial layer of air above freezing temperature lying over a layer of air below freezing point. Under

A

B

C

Plate 92

A. Breakage of branches on red pine due to excessive weight from snow load.

B. Tree bending and breakage as the result of a severe ice storm.

C. Damage to trunks of hardwoods by grinding river ice.

these conditions, precipitation becomes supercooled during its passage through the lower cold layer. On touching any solid object, the supercooled droplets of rain turn instantly to ice. A heavy load of ice bends and breaks branches.

Snow and ice damage to small ornamentals or high-valued trees and shrubs can be prevented or significantly reduced by avoiding planting in exposed sites, providing shelterbelts, and loosely wrapping burlap or plastic mesh around susceptible trees in the winter. Wooden shelters are also effective and are commonly used in urban areas. Plastic bags are not suitable as covers, as relatively high temperatures build up inside them. Even well-vented bags are not recommended. If damage occurs, broken branches should be pruned carefully.

Selected bibliography

Boyce, J.S. 1961. Forest pathology. 3rd ed. McGraw-Hill Book Co., New York, NY. 572 p.
Silverborg, S.B.; Gilbertson, R.L. 1962. Tree diseases in New York State plantations. A field manual. State Univ. Coll. For., Syracuse, NY. Bull. No. 44. 61 p.

Prepared by Pritam Singh.

93. **Wind**
Plate 93

Strong winds are common throughout eastern Canada and are one of the most important physical factors causing damage to trees. In addition to typical strong winds, severe storms, tornadoes, and hurricanes also occur. Damage to both man and the forest from tornadoes and hurricanes can be catastrophic.

Wind damage includes premature defoliation, permanent or temporary bending, breakage of branches and stems, lashing, and even uprooting of trees. Wind also causes shakes or splits in heartwood and damage to roots and root collar by wind rock. Strong prevailing winds can cause deformation, particularly on exposed sites, at high elevations, and on sea coasts. They can also reduce growth in height, thus reducing yield and significantly lowering the aesthetic value of affected trees.

The economic impact of extensive wind damage in the forest can be reduced through salvage operations. Salvage must be started immediately, as insects and fungi attack dead material quickly and lower the quality of the product. Salvage also has the advantage of reducing possible fire hazard and removing material that would support rapid buildup of insect populations.

Wind damage can be mitigated to some extent through the use of shelterbelts, early thinnings, and short rotations. Other silvicultural techniques are available, and their use varies with the circumstances. In urban environments and wooded areas developed for public use, the landowner must be aware of possible damage to property or injury to people from falling trees or branches. Hazardous trees must be identified and corrective measures exercised.

Selected bibliography

Peace, T.R. 1962. Pathology of trees and shrubs with special reference to Britain. Oxford University Press, London. 753 p.
Stone, E.L. 1977. Abrasion of tree roots by rock during wind stress. For. Sci. 23:333-336.

Prepared by Pritam Singh and D.T. Myren.

A

B

C

D

Plate 93

A. A windthrown balsam fir with roots and soil pulled out of the ground.

B. Breakage of balsam fir due to high winds and stem decay.

C. Blow-down caused by high winds in a forested area.

D. Wind breakage of a sugar maple and a largetooth aspen planted as ornamentals.

94. **Winter drying**
Plate 94

Winter drying is a common cause of tree damage in eastern Canada. It affects primarily conifers and can reduce their merchantability as Christmas trees, reduce their value as ornamentals, or even result in tree death. The damage is characterized by browning of the needles above the snow line, which usually becomes visible in the late winter. These needles eventually drop. That portion of the crown protected by snow cover remains green. If winter drying is severe, buds can be killed, but this is a rare occurrence.

Winter drying occurs during midwinter or early spring if cold weather is interrupted by sunny days with slightly higher temperatures and drying winds. These conditions cause transpiration by conifers, and the water lost cannot be replaced because the moisture in the soil is frozen and not available to the roots. It is also possible that water in the main stem is frozen, and this would block upward movement of water. The result is desiccation. The severity of the damage depends upon the frequency and length of exposure to warm temperatures and winds. The damage is worse during winters with little snow, as the soil freezes to a greater depth and more foliage is exposed.

There are some measures the homeowner can take to reduce winter drying. Watering ornamentals regularly during the fall to ensure an adequate moisture supply before winter begins helps, as does mulching to reduce the depth of freezing. Small trees or shrubs should be protected by loose wrappings of burlap in winter. Plastic bags are not suitable as covers, as relatively high temperatures build up inside them. Even well-vented bags are not recommended.

Selected bibliography

Boyce, J.S. 1961. Forest pathology. 3rd ed. McGraw-Hill Book Co., New York, NY. 572 p.

Ostry, M.E.; Nichols, T.H. 1978. How to identify and control noninfectious diseases of trees. U.S. Dep. Agric., For. Serv., North Central For. Exp. Stn., St. Paul, MN. 16 p.

Prepared by Pritam Singh.

Plate 94

Winter drying in a Scots pine plantation. Note the damage is only to the foliage that was above the snow line.

95. **Strangling**
Plate 95

A wire, rope, or string tied tightly and completely around a tree trunk or branch can halt or reduce the translocation of food materials and may result in the death of that portion of the host distal to the point of injury. Cutting a ring of bark around a tree produces a similar result. The material placed around the tree may be tied loosely at first, but the band becomes progressively tighter as the tree grows, until total strangling and death occur.

Pet ropes or chains should be tied to stakes or other nonliving objects. Pets often wrap a tie around a tree, causing injury to the bark. Clotheslines can be fastened to trees by large eyescrews, as can other wires, ropes, and so forth.

Prepared by D.T. Myren and L.P. Magasi.

Plate 95

Strangling of a jack pine by a wire. The portion of the tree above the wire was dead.

96. **Mechanical damage**
Plate 96

Machines such as skidders, bulldozers, trucks, and automobiles cause wounds through collisions with or scraping of trees. The wounds created may prove fatal or may produce an opening allowing secondary pests access to the tree.

Roots of trees require water and air as well as soil nutrients for normal development and to provide food materials to the leaves. When air and water are prevented from reaching the root zone, the tree suffers and may perish. Two problems frequently restrict the penetration of water and air: paving of areas overlaying tree roots, and compaction of the soil by heavy equipment. In construction sites, machines not only compact the soil but also cover the roots with fill as the soil from excavations is spread over the area. The addition of fill alone can cause suffocation. Flooding forested land by impeding drainage causes suffoca-

A

B

C

D

E

Plate 96

A. Damage resulting in dieback and eventual death of white birch as a result of home construction. Addition of fill to the area around the roots and compaction would be major contributors to the problem.

B. Damage to sugar maple by the construction of a ditch alongside the road, which resulted in excessive root loss and injury.

C. Old lawn mower injury on sugar maple. Injury to small trees may cause mortality.

D. Large canker on a hardwood tree caused by an undetermined mechanical factor. Note a canker on a second tree in the background.

E. Mechanical damage to the bark of red pine as a result of a logging operation.

tion of the roots and consequent tree mortality. Man and animals can cause soil compaction, which can be a serious problem in heavily used campgrounds, playgrounds, or pastures.

Smaller machines such as lawn mowers and bicycles can cause tree wounds and, although usually less severe than those caused by heavy equipment, can be a primary cause of mortality. Small wounds have frequently been found to be responsible for allowing entry of fungi into the tree; these fungi can kill the bark, resulting in tree death. Often the first symptom seen is the discoloration or loss of leaves, but a close examination of the affected branches or main stem may reveal an old wound and an area of dead bark. This is often seen near the ground line when mowers are involved.

Grinding by river ice, cuts from animal claws, ice hurled by snow removal equipment, children climbing trees, and a number of other common and even bizarre agents and activities can cause wounds. If pathogenic fungi become established in these wounds, they can kill the bark or invade vascular tissue and bring about the death of the host. In other cases, a wound may allow decay fungi to become established in the tree and, although not necessarily fatal, could result in significant volume loss and predispose the tree to windthrow or breakage.

Control of mechanical damage is best accomplished by prevention, which requires a thorough understanding of the factors that can be responsible. Much of the damage resulting from machine operation can be prevented by creating wells around trees, laying aeration and drainage tiles, and preventing injury to tree stems by erecting barriers or fences. A professional arborist should be consulted regarding these operations.

Prevention by pruning or removal of damaged trees is often the only control available. In forested areas where the problem is severe, salvage operations should be considered. In parks, damaged material could be used as firewood or possibly for rough construction.

Selected bibliography

Harris, R.W. 1983. Arboriculture, care of trees, shrubs, and vines in the landscape. Prentice-Hall, Englewood Cliffs, NJ. 688 p.

Nash, R.W.; Stark, D.; Chadwick, J. 1965. The planting and care of shade trees. Maine For. Serv., Augusta, ME. Bull. No. 22. 60 p.

Prepared by D.T. Myren.

Subject Index[1,2]

---

[1] Major references consulted for Latin binomials and common names:

Farr, D.F.; Bills, G.F.; Chamuris, G.P.; Rossman, A.Y. 1989. Fungi on plants and plant products in the United States. Am. Phytopathol. Soc., APS Press, St. Paul, MN. 1252 p.

Fernald, M.L. 1950. Gray's manual of botany. 8th ed. Illustrated. American Book Co., New York, NY. 1632 p.

Ginns, J. 1986. Compendium of plant diseases and decay fungi in Canada 1960-1980. Agric. Can., Res. Branch. Publ. No. 1813.

Gleason, H.A. 1968. The new Britton and Brown illustrated flora of the Northeastern United States and adjacent Canada. Vols. 1-3. Hafner Publ. Co., New York, NY.

Hosie, R.C. 1979. Native trees of Canada. 8th ed. Fitzhenry & Whiteside Ltd. in cooperation with the Canadian Forestry Service and the Canadian Government Publishing Centre, Supply and Services. 380 p.

[2] The authors for the Latin binomials of causal organisms appear in the text.